# FOOD, INC.

participant°
MEDIA

# FOOD, INC.

**How Industrial Food Is Making Us
Sicker, Fatter, and Poorer—
And What You Can Do About It**

EDITED BY
KARL WEBER

PublicAffairs
*New York*

PublicAffairs books are available at special discounts for bulk purchases in the U.S. by corporations, institutions, and other organizations. For more information, please contact the Special Markets Department at the Perseus Books Group, 2300 Chestnut Street, Suite 200, Philadelphia, PA 19103, call (800) 810-4145 x5000, or email special.markets@perseusbooks.com.

Designed by Trish Wilkinson
Set in 11 point Minion

Library of Congress Cataloging-in-Publication Data
    Food, Inc. : how industrial food is making us sicker, fatter and poorer; and what you can do about it : a participant guide / edited by Karl Weber. — 1st ed.
        p.    cm.
    Includes index.
    ISBN 978-1-58648-694-5 (pbk.)
    1. Food industry and trade—United States. I. Weber, Karl, 1953–
HD9005.F6582 2009
338.4'766400973—dc22                                               2008055405
                            First Edition

                    10 9 8 7 6 5 4 3 2 1

# CONTENTS

# PREFACE

How much do we really know about the food we buy at our local supermarkets and serve to our families? That's the question that filmmaker Robert Kenner sought to answer in *Food, Inc.*

The result is a stunning, mind-expanding movie that lifts the veil on our nation's food industry, exposing the highly mechanized underbelly that's been hidden from the American consumer with the consent of our government's regulatory agencies, the U.S. Department of Agriculture (USDA) and the Food and Drug Administration (FDA). As Kenner shows, our nation's food supply is now controlled by a handful of corporations that often put profit ahead of consumer health, the livelihood of the American farmer, the safety of workers, and our natural environment.

As a result of this truly revolutionary reshaping of the national food supply—caused by business and political trends of just the last forty years—we now have bigger-breasted chickens, the "perfect" pork chop, insecticide-resistant soybean seeds, and even tomatoes that won't go bad during a long trip from the fields to the supermarket shelf. But we also have new, resistant strains of E. coli, the harmful bacteria that causes illness for an estimated 73,000 Americans annually.

We suffer from widespread obesity, particularly among children, and epidemic levels of diabetes and other diet-related illnesses. We have a vast and growing population of farm workers and food processing employees who are underpaid, lack health insurance, and in some cases labor in virtual slavery. We have ugly, foul-smelling factory farms that pollute the air and water while producing foods of dubious safety and nutritional value. We have secretive corporations that increasingly control not just our food supply but the very genetic makeup of the plants that sustain life on Earth. And we have a worldwide economic system that impoverishes farmers in the developing world even as it drives up food prices for the poorest of the poor.

Featuring interviews with such experts as Eric Schlosser (*Fast Food Nation*) and Michael Pollan (*The Omnivore's Dilemma*), along with forward-thinking entrepreneurs like Stonyfield's Gary Hirshberg and Polyface Farm's Joel Salatin, *Food, Inc.* reveals surprising and often shocking truths—about what we eat and how it's produced, but also about who we have become as a nation and where we are going from here.

Because *Food, Inc.* deals with a topic of such importance, complexity, and inherent interest, it's a perfect opportunity to launch the series of film-to-page companion books whose first offspring you now hold in your hand. The American food production system and its impact on our health, our economy, our natural environment, and even our freedoms as a people is a theme with vast ramifications. Understanding it requires drawing a host of connections that no single movie could hope to trace. Hence this book, which is designed to help you take your knowledge about today's food crisis—and your ability to help find solutions—to the next level.

In his chapter on the making of *Food, Inc.*, director Robby Kenner describes the personal journey of discovery he experienced in researching and creating the film. For me, bringing together the contents of this book has been an equally eye-opening journey.

One extraordinary stage in that journey took place over Labor Day weekend in 2008, when my wife Mary-Jo and I attended the Slow Food Nation conference in San Francisco—the first national gathering of Slow Food USA, the American branch of an international organization founded by the Italian cultural critic Carlo Petrini.

Slow Food's original intention was, as the name implies, to combat the spread of American-style fast food and to defend more traditional forms of agriculture and food preparation. It has spread to the United States (as well as around the world) and has now become a popular movement that strives to address and link an array of economic, cultural, and political issues related to the production, sale, and use of food.

After spending four days joining some 85,000 participants in sampling many of the activities offered—including food tastings and sales, panel discussions, film screenings, educational exhibits, and (of course) some amazing dinners—I came away convinced I'd witnessed one stage in the emergence of a new social, political, and economic movement.

The people and organizations who attended the conference came from many varied backgrounds and brought a wide range of interests and values to the table. Some were food lovers for whom the pleasure of fresh, local, well-prepared farm products is the chief motivating factor. Others were economists focused on issues like global hunger and the exploitation of farm workers. And still others were scientists and activists concerned with nutrition, food safety, pollution, and global climate change. Several of the distinguished experts and brilliant writers who would ultimately contribute to this book, from Eric Schlosser and Michael Pollan to Marion Nestle, were in attendance.

The San Francisco gathering convinced me that something big is happening in America today, represented not just by the tens of thousands of people who attended the conference but also by millions of other people around the country who are engaged in similar activities: shopping at organic food stores, at local farmers' markets, or through community-supported agriculture programs (CSAs); ordering fair-trade coffee when they get their morning caffeine fix; asking their kids' schools to get junk food out of the cafeterias; planting community gardens; and writing their representatives to call for changes in farm subsidies, better regulation of meat production, and clearer food labeling standards.

Thanks to concerned Americans such as these, food-related issues—hunger, childhood obesity, rising food prices, water shortages, soil depletion, and many others—are finally achieving a critical mass of attention from the media and the general public. And President Barack Obama has indicated his sympathy for many of the goals of the movement—although, of course, translating that sympathy into concrete reforms against the wishes of a deeply entrenched power base that supports and profits from the current system of industrialized food production will be enormously challenging.

Events such as the release of *Food, Inc.* can play an important catalytic role by bringing together thousands of people and getting them to draw the lines connecting seemingly unrelated economic, political, and social issues. We're hoping the book in your hand will also play a significant supporting role in that consciousness-raising process.

The book is divided into three sections. Part I focuses on the film it-self and includes director Robby Kenner's personal account of the mak-ing of *Food, Inc.* and an interview with co-producer Eric Schlosser, who puts the movie in the broader social context of today's burgeoning food-reform movement.

Part II takes a closer look at many of the issues raised by the movie, providing those who've seen the film (and those who haven't) with much more information about the scientific, economic, political, social, and personal conflicts underpinning the current battle for control of Amer-ica's food supply. Some of the topics discussed in Part II—like the effort by companies such as Monsanto to take control of the genetic basis of our food supply, analyzed by veteran science journalist Peter Pringle, or the appalling conditions suffered by agricultural workers, described by United Farm Workers' (UFW) leader Arturo Rodriguez—will be familiar from *Food, Inc.* Others—like the impact of farming on global climate change, discussed by Anna Lappé, or the effects of the American food system on poor people around the world, dissected by Nobel laureate Muhammad Yunus—are touched on only briefly, or not at all, in the movie. Taken together, the chapters in this part of the book should deepen your appreciation for the importance of the issues treated in the film and the complexities of their interrelationships.

In Part III, we offer some solutions—real-world steps that you can take as a consumer and a citizen to promote improvements in the ways we produce, distribute, and eat food. The advice in this section ranges from the inspirational, as offered up by writer Michael Pollan and eco-agricultural firebrand Joel Salatin; to the intensely practical, including, for example, simple ways to separate food facts from food myths when planning your own family's diet, provided by nutrition expert Marion Nestle; and some ideas on how you can launch a campaign to improve the food served in your neighborhood schools from the citizen-activists at the Center for Science in the Public Interest.

Interspersed throughout the book, among the twelve full-length chapters provided by our distinguished experts, we've included shorter excerpts under the heading of "Another Take." These offer different per-spectives or useful ideas from some of the leading organizations working to improve our ways of producing and eating food, ranging from the

Humane Society of the United States and the Robert Wood Johnson Foundation to Sustainable Table and Food & Water Watch.

Finally, the book concludes with a list of many more resources you can turn to for more information, ideas, and inspiration. They include some of the best books in the field as well as contact information for organizations that are at the forefront of the change effort and URLs for websites that are filled with useful facts related to every major food issue.

One important note: although all of the distinguished individuals and organizations that contributed to this book share a concern about the problems with our industrialized food system and a desire to reform it, they don't always agree on specific details or policy prescriptions. So the appearance of a particular writer or group in these pages shouldn't be construed as an endorsement of every opinion presented elsewhere in the book. Among other things, this book reflects some of the lively ongoing debates among the "food community" about the best directions for the future—debates that, we believe, embody the best American traditions and the hope for a better tomorrow shaped by the contributions of all of us.

Filmmaker Robby Kenner, his colleagues at Participant Media, their publishing partners at PublicAffairs, and all the gifted writers and researchers who helped create the contents of this volume surely join me in hoping that this book, and the movie that inspired it, will become simply the beginning—not the end—of your engagement in the global initiative to create a healthier, more sustainable food supply system for all the peoples of the world.

*Karl Weber*
*Irvington, New York*

# Part I | *FOOD, INC.:* THE FILM

## participant°
MEDIA

*Participant Media & River Road Entertainment Present*
**a Film by Robert Kenner**

# FOOD, INC.

| | |
|---|---|
| Director of Photography | Richard Pearce |
| Coproducers | Eric Schlosser |
| | Melissa Robledo |
| | Richard Pearce |
| Music | Mark Adler |
| Editor | Kim Roberts |
| Executive Producers | William Pohlad |
| | Jeff Skoll |
| | Robin Schorr |
| | Diane Weyermann |
| Producers | Robert Kenner |
| | Elise Pearlstein |
| Director | Robert Kenner |
| Consultant | Michael Pollan |
| Writers | Robert Kenner |
| | Elise Pearlstein |
| | Kim Roberts |
| Post Production Supervisor | Melissa Robledo |
| Main Title & Graphics | Bigstar |
| Associate Producers | Sascha Goldhor |
| | Jay Redmond |

# ONE

# REFORMING FAST FOOD NATION
## *A CONVERSATION WITH ERIC SCHLOSSER*

---

Eric Schlosser is an award-winning investigative journalist whose work has appeared in the *Atlantic Monthly*, *Rolling Stone*, the *Nation*, and the *New Yorker*. His writing has focused mainly on groups at the margins of American society: illegal immigrants, migrant farm workers, prisoners, and the victims of crime. His first book, *Fast Food Nation: The Dark Side of the All-American Meal* (2001), was an international best seller, translated into twenty languages. His second book, *Reefer Madness: Sex, Drugs, and Cheap Labor in the American Black Market* (2003), explored the underground economy of the United States. In *Chew on This* (2006), Schlosser and his coauthor Charles Wilson introduced young readers to many of the issues and problems arising from industrial food production. Two of Schlosser's plays, *Americans* (2003) and *We the People* (2007), have been produced in London. He served as a cowriter and executive producer of the film *Fast Food Nation* (2006). He also served as an executive producer of the film *There Will Be Blood* (2007). For many years, Schlosser has been researching a book about the American prison system.

---

*Q. Your book* Fast Food Nation *was one of the landmarks in the development of today's movement to reform the American food production system. Can you talk about how you got involved as a journalist with issues surrounding food, and how* Fast Food Nation *came to be?*

I was introduced to the world of modern food production in the mid-1990s while researching an article about California's strawberry industry for the *Atlantic Monthly*. It was an investigative piece about illegal immigrants, the transformation of California agriculture, the exploitation of poor migrant workers. It opened my eyes to the difference between what you see in the supermarket and what you see in the fields—the reality of how our food is produced.

So my interest in the whole subject began from the workers' perspective. At the time, the governor of California, Pete Wilson, was arguing that illegal immigrants were welfare cheats. He claimed they were coming to California to live off taxpayers. Instinctively, that didn't sound right to me. During my visits to California, I noticed there were a lot of poor Latinos working very hard at jobs that nobody else seemed willing to do.

The discrepancy between the governor's rhetoric and what I saw with my own eyes made me curious about the actual economic effect of all these illegal immigrants in California. So I began to investigate the subject. And I found that during the same years in which illegal immigration to California had increased, the number of farm workers there had grown too. In fact, California was becoming increasingly dependent on poor farm workers to pick its fruits and vegetables by hand. And lo and behold, some of Pete Wilson's largest campaign supporters were California growers who were profiting enormously from the exploitation of illegal immigrants.

*Q. Purely a coincidence, I'm sure.*

Of course.

Now, I didn't really want to write a political piece about Pete Wilson and why he was such a hypocrite. I've tried hard to avoid writing about politics and politicians. But I wanted to show people that, far from being parasites, these illegal immigrants were propping up California's agricultural industry—which to this day is the most important sector of the state's economy. I had no idea that agriculture was still so important there. When you think of the California economy, you think of high-tech industries like Silicon Valley, you think of Hollywood. You don't think of poor, desperate migrants picking fruits and vegetables with their bare hands. But at the heart of the state's economy is this hard, ugly reality. That was true back in the 1990s, and it's still true today.

So in my article for the *Atlantic*, I wanted to write about farm labor economics, the history of illegal immigration, and the role of illegals in the California economy. But I wanted to do all this by telling the story of something very simple and concrete that we all like to eat: strawberries.

You know, I love strawberries. But when most people see a display of strawberries in their local supermarket, they don't realize that every one of those strawberries has to be very carefully picked by hand. Strawberries are very fragile and easily bruised. So if you want to produce a lot of strawberries in California, you need a lot of hands to pick them. And during the past thirty years—which is the period when, surprisingly enough, the California strawberry became enormous—those hands have belonged to people who are likely to be in the state illegally, who are willing to work for substandard wages in terrible conditions.

Instead of writing a political rant about immigration policy or Pete Wilson, I just wrote something that said, "Look, here's where your strawberries come from—and here's what the consequences are."

That article about migrants in the *Atlantic Monthly* was read by the editors at *Rolling Stone*—Jann Wenner, Bob Love, and Will Dana. They called me into their office and said, "We loved your article, and we'd like you to do for fast food what you did for the strawberry. We want you to write an investigative piece about the fast-food industry. And we want you to call it 'Fast Food Nation.'"

In retrospect, that was a damn good idea. But at the time, I wasn't so sure about it. The editors at *Rolling Stone* didn't know much about the fast-food industry, and neither did I. It wasn't at all clear what the scope or the focus of the article would be. And I didn't want to write something that was snobby and elitist, you know, a put-down of Americans and of their plastic fast-food culture. I still ate at McDonald's then, especially when I was on the road. I really like hamburgers and French fries, and I don't consider myself some kind of gourmand. So I knew what I *didn't* want the article to be, but I wasn't really sure what it should or could be. There was a basic question that needed to be answered: what's the story here?

Jann and Will were really curious about the industry and thought it was worth exploring. So I told them, "Let me think about it."

*Q. By the way, do fast-food companies advertise in* Rolling Stone?

Yeah, the magazine's main readers—young males—are a major demographic for the fast-food chains. Jann Wenner was willing to go after some of his own advertisers, which I give him a lot of credit for.

At first, I wasn't sure whether or not I wanted to accept the assignment. It ran the risk of becoming something terribly kitschy and ironic. So I did what I always do when I want to learn more about a subject: I went to the New York Public Library. Almost everything I write begins at a library— and that is still true today, even with the incredible amount of information available on the Internet. I started reading books about industrial food production and the fast-food industry. Some of the most interesting were memoirs written by the founders of the industry, people like Ray Kroc of McDonald's, Colonel Sanders of Kentucky Fried Chicken, and Tom Monaghan of Domino's Pizza.

I was pretty amazed by what I learned. I was amazed by the size and power of the fast-food industry, by the speed at which it had grown. There was so much that I'd never thought about, like the impact of McDonald's on American agriculture, the role of fast-food marketing in changing the American diet, the obesity epidemic among American children, the huge political and economic influence of the big agribusiness firms.

I was intrigued. So I went back to Will and Jann, and I said, "Yeah, I'll take the assignment. But let's be clear about the scope of this story. I think it's going to lead in all sorts of directions, into all kinds of tangents. This industry has had an impact on many aspects of American society. And I should try to follow the story wherever it leads." And they said, "Great, go for it." So I did.

Researching and writing the article wound up taking me about a year, a lot longer than I thought it would. In the fall of 1998, *Rolling Stone* ran it in two parts. And looking back, although we called the article "Fast Food Nation," it was really never about fast food. It was about this country— about what our food system reveals about our society.

*Q. Are you saying that your work was driven by a political agenda?*

No, I'm much more interested in history and culture and economics than in politics. I don't write with a specific "political agenda" in mind. I try to write things that are complex, that are open to different interpretations, that respect the reader's intelligence. I try to avoid simplistic explanations or ten-point manifestos. The writers whom I've admired most, the ones who have inspired me most, threw themselves into the big issues of their day. They didn't play it safe, hold back, or write for the sake of writing. Writers like Upton Sinclair, John Dos Passos, George Orwell, Arthur Miller, Hunter S. Thompson—they were willing to take risks and go against the grain. My writing deals with many subjects that politicians also deal with. But that doesn't mean I'm interested in writing political tracts. For me, the crucial questions have always been: Is this subject important? Is it relevant? Is it meaningful? Is there something new to be said about it? When the answers are yes, I get to work.

Coming of age in the Reagan-Bush era had a big impact on me. For the past thirty years, so much of American society has been driven by selfishness and greed and a lack of compassion for people at the bottom. I've tried hard in my work to question those motives and offer an opposing view. I've tried to expose hypocrisy and corruption. But what I've tried to do, most of all, is simply to understand the times we live in: What is really going on? What are the driving forces behind the changes we're experiencing? How did things get to be this way?

*Q. So how did the two-part article for* Rolling Stone *become the basis for a book?*

After the article came out, it felt like there was still a lot more to say about the subject. There were a number of issues that I wanted to explore in greater depth. So expanding it into a book seemed the natural next step.

I found the process of reporting the article to be deeply moving. I spent a great deal of time in meatpacking communities, which are sad, desperate places. Seeing the abuse of these meatpacking workers really affected me. Meatpacking used to be one of the best-paid jobs in the country. Until the late 1970s, meatpacking workers were like auto workers. They had well-paid union jobs. They earned good wages, before the fast-food companies came along. It upset me to find that the wages of meatpacking workers had recently been slashed, that they were now suffering all kinds of job-related injuries without being properly compensated.

One of the more remarkable moments of my research occurred while I was visiting a home in the Midwest where a group of impoverished meatpacking workers lived. They were all illegal immigrants. And while I was talking with them, I learned that some of them had worked at a

strawberry farm I'd visited for the *Atlantic Monthly* piece. That's when I
realized that this was a really important story, one that deserved a lot
more of my time and attention. California has been exploiting migrant
workers from Mexico for a hundred years. But that form of exploitation
had, until recently, been limited to California and a handful of South-
western states. Now it seemed to be spreading throughout the United
States. Finding that illegal immigrants were being exploited in the heart-
land of America, in a little town that on the surface looked straight out
of a Norman Rockwell postcard—well, to me, this was something new, a
disturbing and important new trend.

*Q. How much resistance did you encounter in researching and reporting
the book?*

A lot. None of the major meatpacking companies allowed me to visit
their facilities. McDonald's was not at all helpful. The industry, on the
whole, didn't roll out a welcome mat. But many of the workers at fast-
food restaurants and meatpacking plants were eager to talk with me.
They felt that their stories had not yet been told, and they wanted the
world to know what was happening. Their help made *Fast Food Nation*
possible.

Robby Kenner, the director of *Food, Inc.*, has said that his film is not
just about food. It's also about threats to the First Amendment and the
desire of some powerful corporations to suppress the truth. I would agree
with that description of his film, and it also applies to my book. Both of
us, while investigating America's industrial food system, were struck by
the corrupting influence of centralized power. Whenever power is concentrated and unaccountable—whether it's corporate power, governmental power, or religious power—it inevitably leads to abuses. Human
beings are imperfect, and you need a system of checks and balances to
keep them in line, to encourage good behavior. You need competing centers of power. That's not a new idea. That's an old-fashioned American
idea. You can thank James Madison and the other founding fathers for
coming up with it.

And of course this matters a great deal when you're talking about the
food industry. I think the food industry is, by far, the most important industry in every society. Without it, you can't have any other industry. All

the others depend on people being able to eat. It's one thing if competition is eliminated in the baseball card business. That wouldn't be good, but it wouldn't be the end of the world, either, unless you're a baseball card fanatic. When you talk about the food industry, however, you're talking about something fundamental. You're talking about an industry whose business practices help determine the health of the customers who eat its products, the health of the workers who make its products, the health of the environment, animal welfare, and so much more. The nation's system of food production—and who controls it—has a profound impact on society.

Here's an example. One of the major themes of *Fast Food Nation* and *Food, Inc.* is the power of corporations to influence government policy. Again and again, we see these companies seeking deregulation—and government subsidies. They hate government regulations that protect workers and consumers but love to receive taxpayer money. That theme has implications far beyond the food industry. The same kind of short-sighted greed that has threatened food safety and worker safety for years now threatens the entire economy of the United States. You can't separate the deregulation of the food industry from the deregulation of our financial markets. Both were driven by the same mindset. And now we find ourselves on the brink of a worldwide economic meltdown. But in times of crisis we are more likely to see things clearly, to recognize that many of the problems in our society are interconnected. The same guys who would sell you contaminated meat would no doubt sell you toxic mortgages, just to make an extra buck.

One of my goals in *Fast Food Nation* was to make connections between things that might not obviously seem linked. And that posed one of the biggest challenges in writing the book: how far could I go, off on a tangent, before losing readers? I was constantly worried about straying too far and writing something that seemed slightly crazy; I wanted to show the power and influence of this one industry, without exaggerating and suggesting that it somehow ruled the world. There's a fine line between being iconoclastic and being nuts. But it was important to trace the various interconnections. So I wrote about Walt Disney in a book about fast food because Disney greatly influenced how McDonald's marketed its food to children—and that helped change the health of children throughout the world. Some of the things that I learned were truly bizarre, like the fact that Heinz Haber, one of Disney's principal scientific advisers, had been involved with medical experiments performed on concentration camp victims in Nazi Germany. Haber later hosted a Disney documentary singing the praises of nuclear power: "Our Friend the Atom." That fact seemed incredibly bizarre—and yet on some level it also seemed relevant. It made sense, when you're talking about systems that worship uniformity, conformity, and centralized control.

Again and again, I was amazed by where the research led. So much of what I learned seemed incredible. I found myself thinking, "I'm a pretty well-informed person. Why didn't I know about any of this? Why didn't I know about the transformation of the meatpacking industry? Or about the hardships of a new, low-wage workforce? Or about the growth of a flavor industry in New Jersey that invents the taste of almost everything we eat?"

When I spoke to my journalist friends in New York about what I was finding, they didn't know anything about it either. These subjects were not part of the mainstream media's conversation during the late 1990s. I felt like there was this whole world, behind the counter, that had been deliberately hidden from us.

In retrospect, I wish that I'd been able to take a college course on modern food production, like the kind that Michael Pollan now teaches at Berkeley. A decade ago, I was remarkably ignorant about the subject, and it was hard to find information in one place. I pretty much educated myself—and while I uncovered a lot of interesting stuff, there were also some large gaps in my education. After *Fast Food Nation* was published, I met a number of people who had been wrestling with many of the same issues during the 1970s: Wendell Berry, Alice Waters, Orville Schell, Marion Nestle, Francis Moore Lappé. They were trailblazers in this field, and I tried to honor them in subsequent editions of *Fast Food Nation*. Anyone concerned about sustainable agriculture should get to know their work.

Keep in mind, during the 1990s, most of the issues surrounding industrial food production weren't really being discussed in the mainstream. I am very lucky that my editors recognized the importance of this subject, supported my investigative work, and never asked me to tone it down. At

*Rolling Stone,* Jann Wenner, Bob Love, and Will Dana were terrific. Oprah Winfrey had just been sued by the meatpacking industry, and there was every reason for the magazine to fear a similar lawsuit. And it took a lot of nerve for Eamon Dolan, my editor at Houghton Mifflin, to make *Fast Food Nation* his first book at that publishing house. It could easily have been his last one.

I worked hard to make sure everything in the book was as accurate as could be. I didn't want to get sued—and, as a rule, I like to be right. I hate finding even the tiniest mistake in anything I write. So I hired a fact-checker, Charles Wilson, who'd worked at the *New Yorker.* He challenged every single assertion of fact in the book. And I hired a well-known libel attorney, Ellis Levine. Both of them went over the manuscript with a fine-tooth comb. Most of the changes that I had to make were relatively minor ones, altering a word or two here or there. However, I did have to cut a few pages describing some allegedly fraudulent business practices by one of the big meatpacking companies. I didn't have enough evidence to make the accusation. Since I couldn't prove it, I couldn't include it. It was difficult to get rid of those pages because I knew that company was doing some bad things in violation of the law.

**Q.** Fast Food Nation *became a best seller and one of the most influential books of recent years. How did that happen?*

Well, it was a reminder that the conventional wisdom is usually wrong. The major New York publishers weren't interested in publishing the book. They didn't think anyone would want to read it. They didn't think people would care about these things.

In the end, *Fast Food Nation* was published by one of the last independent publishing houses, Houghton Mifflin, based in Boston. My editor there, Eamon Dolan, felt passionately about the subject, took a big risk signing the book, and supported me from beginning to end. There was never any pressure to play it safe. But there was also no expectation on the part of Houghton Mifflin that *Fast Food Nation* would be a best seller. It was launched with a relatively modest marketing campaign. But independent book stores and public radio stations took an interest in the book, and readers began to discover it on their own—especially young people.

I was forty-one years old when *Fast Food Nation* was published in 2001, and I had no idea who would read it. It was amazing to find that most of the people reading the book were at least half my age. They were college students and high school students who'd never known an America without fast food, who'd grown up in a world saturated with fast-food advertising. I'm old enough to remember when the first McDonald's opened in New York City. That was a big deal; I went there the first week it was open.

You might think that people for whom fast food was a routine part of daily life would be among the least likely to respond to the book. You'd think they'd be the most brainwashed by fast-food marketing. But maybe it all makes perfect sense. Maybe if you've been bombarded with these ads, practically since birth, you're even more curious about an alternative view, about a different reality hidden in plain sight.

Now, of course, I'm especially gratified to have so many young people reading the book because if anyone's going to change our industrial food system, it will be them.

When I look at my generation—basically the Reagan-Bush generation—and compare it with the young people of today, I feel hopeful. Kids today are a lot more interested in social issues than my generation was. Just look at the enthusiasm that Barack Obama has generated. There are tons of young people today who are awake and alive, who are questioning the way things are. And there will soon be more of them, as kids graduating from college have to face the consequences of our recent economic policies.

If *Fast Food Nation* had been published in 1995, I doubt that many people would have read it. I'd like to think that my writing is so eloquent and powerful that it would have found an audience no matter when the book appeared, but I'm not that delusional. The truth is that the book appeared at the right moment—in January 2001, a few weeks after George W. Bush took office, right after the theft of the election. It was a moment when people were suddenly beginning to question what was happening in our country. And some of the problems with our industrial food system were just becoming apparent, as Europe wrestled with an outbreak of mad cow disease. *Fast Food Nation* provided an alternative view of the world. It said that things are not OK, at a time when people were starting to feel that way for a lot of reasons.

Soon, a movement to reform our food system really began to emerge. Parents began to campaign against junk food in schools. Childhood obesity became a national issue. Meat recalls raised questions about food safety. And other books on the subject began to appear. Marion Nestle's *Food Politics* came out in 2002. Michael Pollan's book, *The Omnivore's Dilemma,* followed a few years later. Questions were finally being raised in the mainstream about how we produce and market our food. And the more that consumers learned about industrial agriculture, the angrier they got.

*Q. Now nearly ten years have passed. How has the story of America's relationship to food changed in that time?*

There has been a sea change in American attitudes toward food, especially among the educated and the upper-middle class. And there is now a powerful social movement centered on food. Sustainable agriculture, the obesity epidemic, food safety, illegal immigration, animal welfare, the ethics of marketing to children—all of these things are now being widely discussed and debated. The nature of the discussion isn't always to my liking. But at least the issues aren't being hidden and suppressed. Now they are out in the open.

So there's been a huge change in eating habits and awareness among the well-educated and the upper-middle class. For proof of that, just look at the success of Whole Foods, the Food Network, the rise of celebrity chefs, the spread of farmers' markets, all the best sellers about food. Some people worry that the movement to reform our food system is elitist, that right now it appeals to a narrow segment of society. I think that's a real danger, but you have to keep in mind, lots of social movements started off that way. The abolition movement, the civil rights movement, feminism, environmentalism, they all began among the educated, upper-middle class. My concern about the food movement isn't where it started. I'm much more concerned about where it's headed. *Fast Food Nation* offered a critique of the last thirty years of American history and what happened to ordinary people and the poor during that time. I hope the food movement will grow and extend more broadly throughout society. And we need a government that encourages that. For me, this movement has always been about much bigger issues than

"Does an heirloom tomato taste better than a mass-produced tomato?" I don't see any point in having heirloom, organic tomatoes if they're harvested by slave labor. I want tomatoes that taste good—but I also want tomato pickers to be paid well. Luckily, I think a lot of people are starting to realize that it's possible to have both.

*Q. Do you see actual reform of the food system beginning to occur, beyond such trends as farmers' markets and organic restaurants?*

There's no question that meaningful reform has begun. The Coalition of Immokalee Workers, a wonderful organization that defends the rights of farm workers in Florida, has forged agreements with the leading fast-food chains and with Whole Foods. Organic produce is the fastest-growing and most profitable segment of American agriculture. School districts throughout the country are banning sodas and junk foods. New York City and California have passed menu-labeling laws, and California voters recently backed a referendum on behalf of animal welfare. Everywhere you look, people are changing what they eat and demanding that companies be held accountable for what they sell.

Unfortunately, over the past decade, some things have gotten worse—especially the abuse of meatpacking workers. And food safety has deteriorated significantly, with some of the biggest recalls in U.S. history occurring in the last few years. The administration of President George W. Bush was completely in bed with the large meatpacking and food-processing companies. As a result, food safety regulations were rolled back or ignored. These industries were pretty much allowed to regulate themselves. And tens of thousands of American consumers paid the price, with their health.

The fast-food industry has done some good things in the areas of animal welfare, antibiotic use among livestock, and food safety. But the big chains are pretty much operating the way they always have. They want their products to be cheap and taste everywhere exactly the same. That requires a certain kind of production system, an industrial agriculture responsible for all sorts of harms. And the fast-food chains want their labor to be cheap as well. The fundamental workings of this system haven't changed at all since *Fast Food Nation* was published.

In the next few years, I hope to see the same new awareness about food that you find among the well-to-do—and the same access to good food—spreading among the poor, among ordinary, working people. If that doesn't happen, we will wind up with a society in which the wealthy are eating well and staying fit, and everyone else is eating cheap, crappy food and suffering from poor health. At the moment, about two-thirds of the adult population in the United States is obese or overweight. That's the recipe for a public health disaster, and if the number grows much higher, it will be a monumental disaster.

*Q. Some people blame economics for the bad eating habits a lot of Americans practice. Is it true that healthy eating costs more than unhealthy eating?*

Technically, no. It's possible to go to the market, buy good ingredients, and make yourself a healthy meal for less than it costs to buy a value meal at McDonald's. But most people don't have the time or the skills to do that. It's a hell of a lot easier to buy your meal at the drive-through. I can understand why a single parent, working two jobs, would find it easier to stop at McDonald's with the kids rather than cook something from scratch at home.

But we're looking at the whole economic issue the wrong way. Instead of asking, "What does it cost to eat healthy food?" we should be asking, "What's the real cost of this fast, cheap food?" When you look at the long list of harms, this fast, cheap food is much too expensive.

For example, the Centers for Disease Control and Prevention estimate that one-third of all American children born in the year 2000 will develop diabetes as a result of poor diet and lack of exercise. So when we talk about bringing healthy food to every American—yes, it probably means spending more money on food. But you can spend that extra money on food now, or spend a lot more money later, treating heart disease and diabetes.

The fast-food industry didn't suddenly appear in a vacuum. The industry's growth coincides neatly with a huge decline in the minimum wage, beginning in the late 1960s. When you cut people's wages by as much as forty percent, they *need* cheap food. And the labor policies of the fast-food industry helped drive those wages down. For years, the industry has employed more minimum-wage workers than any other—and has lobbied for lower minimum wages. So we've created a perverse system in which the food is cheap at fast food restaurants because they employ cheap labor, sell products that are heavily subsidized by the government, and sell them to consumers whose wages have been kept low. We're talking about a race to the bottom. We shouldn't have a society where the only food that's readily affordable is unhealthy food.

*Q. So how can we break this cycle?*

Well, we can start by taking care of children in this country, rather than simply talking about "family values." We can invest in bringing healthy food into public schools and teaching children about nutrition—like Alice Waters has done, at the Edible Schoolyard. We can begin to change the food culture of this country by changing how we feed and educate our children.

We can create a health-care system that looks after everybody in the country, rich or poor, that cares more about preventing illness than about medicating it, that intervenes long before people need heart bypass surgery or dialysis.

We can raise wages and remove the unfair obstacles that block unionization among farm workers and restaurant workers.

We can make healthy foods more widely available by supporting farmers' markets and bringing supermarkets into low-income neighborhoods. And we can make industrial fast foods more expensive by insuring that the prices at the counter reflect the true costs to society.

We can pass environmental laws that make factory farms clean up their own waste, animal welfare laws that end unnecessary cruelty, and labeling laws that tell consumers what's in their food. This might drive up the price of meat. It might force some Americans to reduce their meat intake. But that might be a good thing. I still eat meat, I'm not a vegetarian—yet. But do we need to eat a large portion of meat two or three times a day, as many Americans do? I don't think so. And if we eliminate some of the factors that keep the price of meat artificially low, it will improve the health of consumers, livestock, and the land.

We can get rid of government subsidies for factory farms and corporate farms. If the government's going to subsidize any foods, it should be healthy foods: fruits, nuts, and vegetables—not high-fructose corn syrup and corn-based cattle feeds. And we need to support family farmers who have a long-term interest in land stewardship, not corporate farms that view the land as just another commodity to be bought, sold, and exploited.

There's no one thing, no simple cure, that is going to transform the American diet or our industrial food system. It's going to be an enormous task. But as revolutions go, this one will be a real pleasure. Making Americans healthier, providing them food that tastes a lot better—we're not talking about imposing some grim sacrifices. And reforming the food system won't harm the United States economically. The food industry isn't going to go away, there aren't going to be massive job cuts. People will still need to eat, but the money they spend for food—and the money the government spends on food—will wind up in different hands. The United States has incredibly productive farmland for grains, fruits, vegetables, and livestock. A reformed food industry would in many ways be more economically efficient. The obesity epidemic is now costing us about $100 billion a year. The medical costs imposed by the fast-food industry are much larger than its annual profits—except the industry isn't paying those medical bills. Obesity may soon surpass tobacco as the number-one cause of preventable death in the United States. How much is each one of those lives worth?

Today there are a lot of complex problems in the United States that will be very difficult to solve. Reforming the food industry isn't one of them. Companies that sell healthy foods should earn large profits; companies that sell junk food shouldn't.

*Q. Yet your agenda for change is so sweeping that it sounds radical.*

Radical? I think my proposals are pretty conservative. It's the industrial food system that seems radical and completely out-of-keeping with tradition. This country thrived for almost two hundred years without industrial fast foods. There's no reason we can't thrive, once again, without them. The way we produce food today, this giant industrial system, is only about thirty years old. And look at the damage it has already done, in such a brief period of time. For most of our history, we had a very different kind of agricultural system and a very different diet—and that traditional system worked well enough to support a continent full of people, to feed our cities, to help feed the rest of the world.

*Q. So if reforming the food system is actually a feasible project—more a matter of restoration than of radical change—the question is one of public will. How close are we, do you think, to a tipping point?*

I don't know when we'll reach the tipping point. But I do know we're a hell of a lot closer now than we were ten years ago. I'm optimistic.

# Another Take | FOOD SAFETY CONSEQUENCES OF FACTORY FARMS

### By Food & Water Watch

---

Food & Water Watch is a nonprofit consumer organization that works to ensure clean water and safe food. Working with grassroots organizations around the world, it strives to create an economically and environmentally viable future by challenging the corporate control and abuse of our food and water resources, empowering people to take action, and by transforming the public consciousness about what we eat and drink.

In "Food Safety Consequences of Factory Farms," Food & Water Watch provides an overview of some of the devastating effects—on consumers, on the environment, and on animals themselves—of the widespread system of industrialized farming that has become the norm in twenty-first century America. For more information, visit the organization's website at http://www.foodandwaterwatch.org.

---

Today, many of the meat and dairy products sold in the United States come from factory farms—industrial-scale facilities where tens of thousands of animals are crowded together in tight conditions and cannot carry out normal behaviors such as grazing, rooting, and pecking.

The environmental and economic effects of factory farms on rural communities are well known. These facilities cannot process the enormous amounts of waste produced by thousands of animals, so they pour and pile manure into large cesspools and spray it onto the land. This causes health problems for workers and for neighbors. Leaks and spills from manure pools and the runoff from manure sprayed on fields can pollute nearby rivers, streams, and groundwater. And the replacement of independently owned, small family farms by large factory operations often drains the economic health from rural communities. Rather than buying grain, animal feed, and supplies from local farmers and businesses, these factory farms usually turn to the distant corporations with which they're affiliated.

But even if you live in a city hundreds of miles from the nearest factory farm, there are still lots of reasons to be concerned about who is producing—and how—the meat and dairy products you and your family consume.

## ANIMAL FEED—YOU ARE WHAT YOU EAT . . . AND WHAT THEY ATE

Factory farm operators typically manage what animals eat in order to promote their growth and keep the overall costs of production low. However, what animals are fed directly affects the quality and safety of the meat and dairy products we consume.

***Antibiotics.*** Factory farmers typically mix low doses of antibiotics (lower than the amount used to treat an actual disease or infection) into animals' feed and water to promote their growth and to preempt outbreaks of disease in the overcrowded, unsanitary conditions. According to the Union of Concerned Scientists, seventy percent of all antimicrobials used in the United States are fed to livestock.[1] This accounts for twenty-five million pounds of antibiotics annually, more than eight times the amount used to treat disease in humans.[2]

The problem is this creates a major public health issue. Bacteria exposed to continuous, low-level antibiotics can become resistant. They then spawn new bacteria with the antibiotic resistance. For example, almost all strains of staphylococcal (staph) infections in the United States are resistant to penicillin, and many are resistant to newer drugs as well.[3] The American Medical Association, American Public Health Association, and the National Institutes of Health all describe antibiotic resistance as a growing public health concern.[4] European countries that banned the use of antibiotics in animal production have seen a decrease in resistance.[5]

**Mad Cow Disease.**    Animal feed has long been used as a vehicle for disposing of everything from road kill to "offal," such as brains, spinal cords, and intestines. Scientists believe that "mad cow disease," or bovine spongiform encephalopathy (BSE), is spread when cattle eat nervous system tissues, such as the brain and spinal cord, of other infected animals. People who eat such tissue can contract variant Creutzfeldt-Jakob disease (vCJD), which causes dementia and, ultimately, death. Keeping mad cow disease out of the food supply is particularly important because, unlike most other food-borne illnesses, consumers cannot protect themselves by cooking the meat or by any other type of disinfection. The United States has identified three cases of mad cow disease in cattle since December 2003.

In 1997, the Food and Drug Administration (FDA), the agency that regulates animal feed, instituted a "feed ban" to prevent the spread of the disease. Although this ban provides some protections for consumers, it still allows risky practices. For example, factory farm operators still feed "poultry litter" to cattle. Unfortunately, poultry litter, the waste found on the floors of poultry barns, may contain cattle protein because regulations allow for feeding cattle tissue to poultry. And cattle blood can be fed to calves in milk replacer—the formula that most calves receive instead of their mother's milk. Finally, food processing and restaurant "plate waste," which could contain cattle tissue, can still be fed to cattle.

In 2004, after the discovery of BSE in the United States, the FDA had the opportunity to ban these potential sources of the disease from cattle feed. But instead, officials proposed a weaker set of rules that restricted some tissues from older cattle. A safer policy for consumers would be to remove all tissues from all cattle from the animal feed system, regardless of their age, and also to ban plate waste, cattle blood, and poultry litter.

In the fall of 2006, the U.S. Department of Agriculture (USDA) decided to scale back testing for mad cow disease. Officials cited what they claimed was the low level of detection for the disease in the United States. Now, only 40,000 cattle, one-tenth the number tested the year before, will be tested annually. Given the weakness of the rules that are supposed to prevent the spread of the disease, this limited testing program effectively leaves consumers unprotected.

**E. Coli.**   Cattle and other ruminants (animals with hooves) are uniquely suited to eat grass. However, in factory farm feedlots, they eat mostly corn and soybeans for the last few months of their lives. These starchy grains increase their growth rate and make their meat more tender—a process called "finishing." However, scientists point to human health risks associated with the grain-based diet of "modern" cattle.

A researcher from Cornell University found that cattle fed hay for the five days before slaughter had dramatically lower levels of acid-resistant E. coli bacteria in their feces than cattle fed corn or soybeans. E. coli live in cattle's intestinal tract, so feces that escape during slaughter can lead to the bacteria contaminating the meat.[6]

Vegetables can also be contaminated by E. coli if manure is used to fertilize crops without composting it first, or if water used to irrigate or clean the crops contains animal waste. The 2006 case of E. coli–contaminated spinach offers a dramatic example of how animal waste can impact vegetables.

**Fat.**   According to a study by the Union of Concerned Scientists, beef and milk produced from cattle raised entirely on pasture (where they ate only grass) have higher levels of beneficial fats, including omega-3 fatty acids, which may prevent heart disease and strengthen the immune system. The study also found that meat from grass-fed cattle was lower in total fat than meat from feedlot-raised cattle.[7]

## PROMOTING GROWTH AT ANY COST

Factory farms strive to increase the number of animals they raise every year. To do so, however, they use some practices that present health concerns for consumers.

**Hormones.**   With the approval of the FDA and USDA, factory farms in the United States use hormones (and antibiotics, as discussed earlier) to promote growth and milk production in beef and dairy cattle, respectively. Regulations do prohibit the use of hormones in pigs and poultry. Unfortunately, this restriction doesn't apply to antibiotic use in these animals.

An estimated two-thirds of all U.S. cattle raised for slaughter are injected with growth hormones.[8] Six different hormones are used on beef cattle, three of which occur naturally, and three of which are synthetic.[9] Beef hormones have been banned in the European Union since the 1980s. The European Commission appointed a committee to study their safety for humans. Its 1999 report found that residues in meat from injected animals could affect the hormonal balance of humans, causing reproductive issues and breast, prostate, or colon cancer. The European Union has prohibited the import of all beef treated with hormones, which means it does not accept any U.S. beef.[10]

Recombinant bovine growth hormone (rBGH) is a genetically engineered, artificial growth hormone injected into dairy cattle to increase their milk production by anywhere from eight to seventeen percent.[11] The FDA approved rBGH in 1993, based solely on an unpublished study submitted by Monsanto.[12] Canada, Australia, Japan, and the European Union all have prohibited the use of rBGH.

Approximately twenty-two percent of all dairy cows in the United States are injected with the hormone, but fifty-four percent of large herds (500 animals or more), such as those found on factory farms, use rBGH.[13] Its use has increased bacterial udder infections in cows by twenty-five percent, thereby increasing the need for antibiotics to treat the infections.[14]

In addition, the milk from cows injected with rBGH has higher levels of another hormone called insulin-like growth factor-1 (IGF-1). Elevated levels of IGF-1 in humans have been linked to colon and breast cancer.[15] Researchers believe there may be an association between the increase in twin births over the past thirty years and elevated levels of IGF-1 in humans.[16]

## UNWHOLESOME, UNSANITARY, AND INHUMANE CONDITIONS

Raising animals on cramped, filthy, and inhumane factory farms differs greatly from what most consumers envision as the traditional American farm.

*Disease.*    Hundreds of thousands of birds are breathing, urinating, and defecating in the close quarters of factory-style poultry farms. These conditions give viruses and bacteria limitless opportunities to mutate and spread. This is a very real concern given the presence of avian flu in many parts of the world. The poultry industry has tried to portray factory farms as a solution to the spread of avian flu. It claims that keeping the birds indoors somehow isolates them from the outside world and the disease that lurks there.

Contrary to these claims, scientists suspect that it was in poultry factory farms that avian flu mutated from a relatively harmless virus found in wild birds for centuries to the deadly H5N1 strain of the virus that is killing birds and humans today.[17] In England, the virulent H5N1 strain first broke out at the country's largest turkey farm in early 2007. Theories about the source of the infection include rats or flies entering the facility from a nearby poultry processing plant that itself had received a shipment of infected poultry parts from Hungary.[18] These large-scale facilities rely on truckloads of feed and supplies that arrive every day, providing a way for the disease to spread.

**Contamination.**    Raising thousands of animals together in crowded conditions generates lots of manure and urine. For example, a dairy farm with 2,500 cows produces as much waste as a city of 411,000 people.[19] Unlike a city, where human waste ends up at a sewage treatment plant, livestock waste is not treated but rather washes out of the confinement buildings into large cesspools, or lagoons. In feedlots, open lots where thousands of cattle wait and fatten up before slaughter, the animals often stand in their own waste before it is washed away. The cattle often have some water-splashed manure remaining on their hides when they go to slaughter. This presents the risk of contamination of the meat from viruses and bacteria.

## ANIMAL WELFARE

Rather than grazing in green pastures, animals on factory farms exist in tight confinement with thousands of other animals. They have little chance to express their natural behaviors.

Pigs on factory farms are confined in small concrete pens, without bedding or soil or hay for rooting. The stress of being deprived of social interaction causes some pigs to bite the tails off of other pigs. Some factory farm operators respond by cutting off their tails.

Chickens stand in cages or indoors in large pens, packed so tightly together that each chicken gets a space about the size of a sheet of paper to itself. The chickens are not given space to graze and peck at food in the barnyard, so they resort to pecking each other. Many factory farmers cut off their beaks, a painful procedure that makes it difficult for chickens to eat.

## THE TREND CONTINUES:
## FROM FACTORY FARM TO TABLE

Factory farming is but one component of the industrial meat production system. Just as small farms have given way to factory farms, small meat plants are disappearing while large corporate operations have grown even bigger—and faster. While these trends increase production and profits for the industry, they also increase the likelihood of food contamination problems. Although the government provides inspectors to protect consumers, their authority is waning as the government gives greater responsibility to the industry to self-regulate.

## CONSUMERS CAN SAY
## NO TO FACTORY FARMS

*Vote with Your Dollars.*    Know where your meat comes from. Refer to the *Eat Well Guide* to find a farm, store, or restaurant near you that offers sustainably raised meat and dairy products: http://www.eatwellguide.org.

Or buy your meat directly from a farmer at a farmers' market. Talking with the farmers at a farmers' market in person will give you the chance to ask them about the conditions on their farm. You can find farmers markets at http://www.ams.usda.gov/farmersmarkets/map.htm, and you can find questions to ask a farmer at http://www.sustainabletable .org/shop/questions.

Organic meat is also a good choice because the organic label means that the product has met standards about how the meat was produced. Visit our website at www.foodandwaterwatch.org to check out our labeling fact sheet to find out more about which labels to look for. And check out our milk tip sheet to find out which milk labels to look for and our product guide for rBGH-free dairy products in your area.

**TWO**

# EXPLORING THE CORPORATE POWERS BEHIND THE WAY WE EAT
## *THE MAKING OF* FOOD, INC.

### *By Robert Kenner*

---

Award-winning director Robert Kenner completed production on *Food, Inc.* for Participant Media and River Road Entertainment in the fall of 2008.

Previously, Kenner received the 2006 Peabody, the Emmy for exceptional merit in Non-Fiction Film-Making, and the Greirson (British Documentary award) for his film *Two Days in October*. In its review, the *Boston Globe* commented, "If you could watch only one program to grasp what the Vietnam War did to the U.S. . . . *Two Days* would be a great choice. . . . It is profound."

Kenner's other notable work includes his co-filmmaking endeavor on the Martin Scorsese *Blues Series*, which *Newsweek* called "as fine a film [as has] ever been made about American music" and "the unadulterated gem of the Series." Kenner's documentaries for *The American Experience* include *War Letters*, *John Brown's Holy War*, and *Influenza 1918*. He also directed numerous specials for National Geographic, including *Don't Say Goodbye*, which won the International Documentary Award and Cable Ace, Genesis, and Emmy awards.

---

In the making, *Food, Inc.* turned into a movie with a cause—an exposé of sorts about America's industrial food system and the toll it takes on our health, our environment, our economy, and the rights of workers. But it wasn't because I set out to make a polemic about this system. My original concept derived from a curiosity about representing the multiple voices and points of view of the people who bring food to our tables.

I was taken by surprise by how much my original concept changed during the making of the film. It happened because I was repeatedly denied access to the companies I sought to film. I met people who wanted to talk, but who were silenced by fear, who were scared to speak freely, scared to allow me onto their properties, scared to be seen in public with a camera nearby, and even scared to tell me what they eat. I gradually realized that, while I had set out to make a film about food, I was now making a film about unchecked corporate power. The fear that had initially surprised me became a familiar part of my working days and sleepless nights, as I worried about who might come after me for telling this story. I didn't realize when I started that I was really making a film about freedom of speech, and I didn't know that I would spend the next few years of my life working with First Amendment lawyers. If I had known at the outset the challenges of making a film about the food industry, I might have chosen another subject.

―――――――

I like to make films about social issues told through the lens of personal stories. In 2001, I had just finished a film for *The American Experience* called *War Letters*. Eric Schlosser saw it and liked it. Each of us swears he was the first to contact the other, but however it happened, it was clear that we would be making a movie together.

If you know anything about food issues, you know Eric's name. As the author of the longtime best seller *Fast Food Nation*, Eric is one of our leading journalistic voices about the problems of our food industry and a strong advocate of reform. He's a muckraking journalist very much in the mold of such heroes as Sinclair Lewis, Lincoln Steffens, Ida Tarbell, and I. F. Stone—an unabashed advocate of the rights of working people, consumers, and ordinary citizens, one who isn't afraid to take on the powers-that-be, armed only with the truth as he sees it. Like millions of other Americans, I'm a longtime admirer of Eric's writing and a fan of his work.

Eric and I got in touch. A friendship began, and we soon started talking about the possibility of doing a documentary together.

Naturally, we talked about doing a film based on Eric's watershed work, *Fast Food Nation*. But somehow the right combination of circumstances—my availability, Eric's availability, production funding—never came together at the right time, although we came close a number of times. Five years after Eric and I met, we took the proposal to Participant Media, the film company responsible for *An Inconvenient Truth*, *Syriana*, and many other socially conscious films. We brought it to Diane Weyermann and Ricky Strauss, and they liked what they heard.

What emerged from this development process was a thirty-page treatment that drew from stories in *Fast Food Nation* and new stories that Eric and I had discovered. These included, for example, a "flavor factory" that adds the charbroiled taste to burgers that Eric had written about in his book, and a new story about how adolescents are actually given MRIs to measure the ways their brain waves respond to fast-food ads.

This was great material. But during this time, when I told people what I was working on, I felt discouraged because many mistakenly thought that Morgan Spurlock's documentary, *Super Size Me*, was already a documentary version of *Fast Food Nation*. Spurlock's documentary dealt with the negative effects of eating fast food. But in reality this aspect of fast food represented only a small fraction of Eric's groundbreaking look at how the impact of the fast-food industry transformed our entire food system. I still felt there was room for a documentary about the industrialization of our food system.

I also realized that, since Eric's book had used fast food as the lens through which to examine the industrialization of our food system, there were many people who incorrectly thought the book was just about fast food. If they avoided fast food, they thought, they could avoid industrial food. They didn't realize that the hamburger meat and the chicken breasts they were buying in the supermarket come from the same sources as the fast-food meat. And they definitely did not think that the tomatoes and spinach they were buying in the market had anything to do with this industrialized system. In my new picture, I wanted to show why these assumptions were wrong.

Right around this time, Michael Pollan's book *The Omnivore's Dilemma* was published and followed *Fast Food Nation* to the top of the best seller lists. *The Omnivore's Dilemma* dealt with the ethical, cultural, and economic implications of the choices we make about eating. By focusing particularly on the role of government in promoting the industrialization of food production, Michael broadened the theme enormously, making it even more relevant to all Americans than it had already been. Michael had written a number of groundbreaking books that tied the agricultural and cultural worlds together in a fresh and intriguing way. His *Omnivore's Dilemma* led me to reframe our story. I got to know Michael, and he joined the project as another powerful ally and adviser in helping to shape what would become *Food, Inc.*

While Participant Media and I were shopping the treatment around, looking for the other fifty percent of the money we needed for production, I was simultaneously realizing that this film needed to tell a bigger story than the initial version we presented in the treatment.

I remember speaking to a vegetarian movie executive at one of the big studios who assumed that the problems I was describing were of minimal concern to her. When the horrific outbreak of E. coli contamination in packaged spinach hit the headlines, suddenly the executive was on the phone wanting to know more about the food safety issues I'd been researching.

The spinach outbreak was a wake-up call for many people, including me. We suddenly realized that we were all vulnerable. It was the moment when I knew clearly that the film needed to tell the whole story of the food we eat, from the transformed seeds we plant to the food we buy at the supermarket. The entire food system, I was discovering, had been affected by the same forces of efficiency, uniformity, and conformity as the fast-food industry, and yet most people didn't realize it.

I wanted to understand the complexities of trying to feed a vast nation. I was discovering that, on one hand, we were paying less for our food than any people in history, yet on the other I was beginning to get a glimpse at the true price our society was paying for this low-cost food.

The film project became a reality when Participant Media brought in River Road Entertainment to provide the rest of the funding. It was an exciting and terrifying moment. Suddenly I had to actually go make this movie, and I knew that the treatment and the six months of research that went into it were not necessarily applicable to where we were now going. But I was soon to discover that this was the least of my problems.

———————

As a filmmaker, I don't naturally gravitate to big subjects like the food supply. Big subjects are hard to humanize, hard to put in the right scale for viewers to understand, identify with, and fully experience. For this reason, I generally prefer topics that are more self-contained. Nonetheless, the potential for this movie was enormous, especially with Eric and Michael's involvement.

The more I learned about how the world of food had been altered without our knowing about it, the more interested I became in telling the story. I sensed it would be difficult to capture the issue's complexity in a single ninety-five-minute snapshot. There were so many stories to tell of how this low-cost food was taking such a high toll on the workers who made it, the environment that couldn't sustain it, and the health of the consumer who eats it.

I should explain that I came to this topic with very few preconceived notions about food (other than a fascination with its social, cultural, and economic significance derived from books like Eric's and Michael's). I do love good food, which I can attribute largely to the influence of my wife Marguerite, who is a wonderful cook. I am not a purist about healthy or ethical eating. But I now find that the "industrial food" most of us are accustomed to eating doesn't taste quite the same to me as it did before I made the film. Before I began the research process, I was probably a lot like the average person who will watch *Food, Inc.* And I hope that means that the facts I learned about our food system—some of them amazing, some disturbing, and many simply fascinating—will interest moviegoers as much as they interested me.

Our team—producer Elise Pearlstein, coproducer and cinematographer Richard Pearce, and I—set to work. We spent a lot of time learning more about the broad panorama we hoped to depict in the movie. Eric and later Michael introduced us to scores of fascinating people—experts inside and outside the food industry; organic farmers; labor organizers;

advocates for healthy eating; nutritionists and agricultural researchers; and ordinary workers in the food production, processing, and retailing businesses who were willing to serve as sources about what was really going on inside this surprisingly little-known field. Eventually we ended up speaking with hundreds of people, with each conversation deepening our understanding of the complexities of the issues we wanted to address.

I made a couple of basic decisions. One was that I didn't want the movie to lean on the connecting thread of a central narrator (unlike, for example, *An Inconvenient Truth*, in which the personality, career, and passion of Al Gore as a crusader against global warming serves as the movie's driving force). Without a narrator, I faced the challenge of how to convey a tremendous amount of information in a concise and entertaining manner. Utilizing graphics and animation helped solved this problem while adding style and humor to the film.

Another decision was that the picture ought to contain a number of stories that were unfolding in the present, rather than retelling events that had already transpired—brief, somewhat self-contained narratives, filmed in cinema verité style, that could illustrate, personalize, and bring to life the many problems of the food production system. But how would we weave these fragments into a coherent whole? That proved to be the major challenge in making the picture work—one that we've continued to wrestle with throughout the production process.

I had a particular perspective in mind. In the movie, I hoped to tell the story from multiple points of view, presenting not just the critics but those who view the industrial system as the most successful food-production system the world has ever known. I'd used a somewhat similar collage technique in my 2005 *American Experience* documentary *Two Days in October*. In that movie, based on the book *They Marched into Sunlight* by David Maraniss, I'd depicted two closely interwoven, almost simultaneous events from 1967—a campus antiwar demonstration in Madison, Wisconsin, protesting on-campus recruiting by the Dow Chemical company, maker of the defoliant napalm, and a massacre that took place in Vietnam the next day—telling the stories from several points of view. I went into that film concerned that there was not much new to learn about Vietnam. But I ended up learning a great deal from speaking with some amazing soldiers whose attitudes about the war were different from mine, and it helped me rethink my previous understanding of this history.

I wanted to use the same approach here, hoping to get at the complexities and built-in paradoxes of the industrial food supply system—a system that makes affordable, abundant food more readily available than ever before in history. On the other hand, this seems to contribute to a society in which poor nutrition, obesity, and diseases related to poor diet are rampant. This same system has produced enormous wealth for agribusinesses, food processors, and the chemical and technology companies that design our foods, while relying on the work of disenfranchised, low-paid migrant laborers to bring in the harvest and fill with abundance the tables of the affluent.

My goal was to reach a larger audience than just the food activist by including many divergent points of view about this system.

Unfortunately, the industrial voices proved to be difficult to include. And the reason was simple: the food-industry representatives whose voices I hoped to capture in *Food, Inc.* declined to participate.

In the early months of the project, I thought—perhaps naively—that I could work with the food companies. I spoke with industry spokespeople from many of the largest companies in the food business—Smithfield, Tyson, Perdue, Pepsi, General Mills, Monsanto—as well as with advocates from industry organizations representing poultry producers, beef producers, and agribusiness. Many of them were genuinely nice people, and some of them offered us basic information that helped improve our understanding of the work they did.

But before long, we hit a wall. I was shocked to discover that there was an absolute, impenetrable barrier separating the information they were willing to share with me and the moviegoing public and the vast dark areas of their operations that they were determined to keep private. In all my years as a nonfiction moviemaker, I never experienced so many people so consistent in their determination *not* to speak with me—despite the fact that they *knew* my movie would be made, and that only by speaking with me could they ensure that their perspective would be reflected on screen.

I repeatedly tried to get through this wall of silence, which proved quite frustrating. A Monsanto representative agreed to let us film at an industry conference on the development, sale, patenting, and control of seeds—one of the most obscure and important topics we cover in the movie—only to have the convention officials exclude us at the last minute, saying they could only admit journalists who were connected with the food companies. The idea that writers who are paid publicists for corporations should be considered "journalists" struck me as exactly backward.

One of the industrial farmers who actually made it into the picture is Vince, a chicken grower from Kentucky. He raises poultry under a contract with Tyson, very much the same deal that lots of once-independent chicken farmers now operate under. We spent some time with Vince and got to know him fairly well. We discovered that his mom hates the chicken industry—the smells it generates, the pollution it produces, the negative effects it has on the community. But Vince likes it, mainly because it has proven to be profitable for him. (When you watch *Food, Inc.*, you'll see Vince driving up to one of his chicken houses and commenting on the stench in the air: "It smells like money!")

Vince was a good guy, and he taught us a lot about how his business works. But soon after he spoke with us, a team of around ten people from Tyson headquarters showed up at his farm and suggested, indirectly, that he'd be making a big mistake if he let us film there. They never actually *forbade* us to enter Vince's chicken houses. But, as Vince put it, he felt like there was "a big red flag flying," the kind you see at the beach on a stormy day that means, "Swim at your own risk." With his livelihood dependent on Tyson, Vince didn't feel like taking that kind of risk—understandably.

We ran up against the same kind of barrier over and over again, in one industry after another. Everyone was civil to us but also very firm about the limits to their cooperation. And we soon learned that the self-protective mechanisms of the food industry extend far beyond just offering a polite "No" to requests from journalists.

It was incredibly frustrating to be unable to tell the story in the way I had imagined. The more we were unable to tell the story because of lack of access and cooperation, the more it felt like a story that needed to be told. But how could we make the audience aware of what we were experiencing—this iron curtain separating us from seeing where our food comes from? I didn't want to be like Michael Moore pushing my way into General Motors or filming myself getting rejected on the telephone. I felt that style would imply a very different approach than the one I was taking. But at the same time, when people refuse to participate in a public dialogue, how do you let the audience know this, without having a narrator to help explain? We thought about having the names of companies that refused us scroll across the screen. But ultimately we went with matter-of-fact text stating when a company refused to participate. When this is repeated over and over again, the cumulative message to the audience hopefully becomes very clear.

I think the turning point for me in understanding just what we were up against—and what our movie is really all about—came when we were filming the sequence about Barbara Kowalcyk.

As you'll recall from the picture, Barb and her husband lived in Wisconsin with their two-and-a-half-year-old son, Kevin. He was a beautiful, perfectly healthy child, until one day he ate a hamburger containing E. coli bacteria. Kevin developed a complication most people have never ever heard of—hemolytic uremic syndrome—and within twelve days he was dead, one of the thousands of victims of food-borne disease who die in the United States every year. And even after Kevin's death, it took weeks for the tainted meat to be removed from the market—a shocking comment on the failures of our regulatory system.

As we show in the movie, sicknesses from contaminated food are on the rise, largely because of the lobbying power of the big food processors, which has crippled efforts to police the industry effectively or to impose meaningful sanctions on companies guilty of unsafe food-handling practices. Today, Barb has started a nonprofit organization, the Center for Foodborne Illness Research & Prevention, and she devotes her time to advocating on Capitol Hill for better regulation of the food industry.

As you can imagine, when we heard Barb tell her story, we all got pretty emotional. (My coproducer Elise Pearlstein actually had to leave the room during the filming because she couldn't stop crying.) But that wasn't what shocked me. What shocked me was when I asked her how the death of Kevin and her subsequent research into the problems with our food supply have affected her family's current eating habits.

"I can't answer that," Barbara told me, "because if I did, I could face a lawsuit."

This left me speechless. I didn't intend to put my inarticulate responses in the film, but I felt that there was no other way to adequately convey Barb's situation. Barb is no conspiracy-monger or left-wing anti-corporate extremist. She's a lifelong Republican. But here she was, telling me she was actually *afraid* to speak openly about her opinions because of the unchecked power of the food corporations.

It was my first encounter with the so-called "food disparagement" laws (sometimes called "veggie libel" laws), which have been passed in thirteen states. These laws make it easier for food industry companies to sue people who criticize the quality of their products. In some cases they even establish looser standards of proof than apply in other kinds of libel cases. You might remember that Oprah Winfrey was sued under the Texas version of these laws after she hosted a program about bovine spongiform encephalopathy (BSE), or "mad cow disease." She ultimately won that case, but the lawsuit had a chilling effect on many people's willingness to speak out about food safety.

By this point, I knew that the food companies had no interest in talking on camera, but until that conversation with Barbara Kowalcyk, I had no idea how effective they were at intimidating other people and preventing them from talking. And I was increasingly startled and dismayed to hear the same kind of sentiment popping up from other people I spoke with during the filming of *Food, Inc.*—not just in regard to veggie libel, but in regard to almost *any* effort to buck the power of the big food companies. We saw in one story after another how the threat of lawsuits created fear and intimidation that spread through communities. The footage of seed clearer Moe Parr talking about his lawsuit with Monsanto still haunts me to this day—you can see the fear etched in his face.

I began to see the far-reaching consequences of the current state of our food system, in which just a tiny handful of companies dominates, and as a result, these companies wield disproportionate power. Some of them, like Monsanto, now literally own and control life forms, making it increasingly difficult for farmers to grow basic crops unless they pay a fee to the corporation for the privilege. And thanks to the intimate connections these corporations have with government, the agencies that are supposed to oversee their operations and protect the interests of the public are largely toothless. In fact, as we show in the film, many of those agencies are actually managed by former food industry lobbyists, lawyers, and executives—a classic case of the fox guarding the hen house.

I felt as if I'd stumbled into an Orwellian world of behind-the-scenes wire-pullers controlling a fundamental aspect of our lives—the food we eat—which operates in near-total secrecy thanks to the fear it instills in people who know about it. The food companies have a vested interest in controlling what we know—or think we know—about the foods we eat. They desperately want to sustain the myth that our food still comes from a pristine farm with a red barn and a white picket fence rather than the factory farm that really produces it. So the less we think about the reality of our food system, the better for them.

And that's how *Food, Inc.* evolved into something more than a movie about the food industry (important as that subject is). It became a movie about our freedoms as Americans, and about how powerful companies tightly interlocked with government pose a serious threat to those freedoms.

This made the project quite different from any other movie I've worked on. I'd never before spent so much time consulting with lawyers while making a movie. This story was so important and so controversial—and our potential adversaries so threatening—that we went to great lengths to verify the facts. We found ourselves at times deleting information that we believed to be true because we were concerned that we didn't have adequate sources to back it up.

Under the circumstances, I have to say that I'm impressed with the integrity and honesty of the people who consented to appear in the movie. Some of them represent to me what businesspeople should be like. For example, there's Gary Hirshberg, CEO of Stonyfield, who talked with us extensively about his organic yogurt business and the efforts he has been making to bring natural production methods into the mainstream of the American food industry.

Not everyone in the organic movement agrees with Gary's philosophy or his approach to business. When he sold the company he founded to Danone, the French corporation that makes Dannon yogurt and that dominates the yogurt business worldwide, many purists were dismayed. But Gary candidly discussed the pros and cons of corporate control with the kind of frankness that I wish more people in the food industry would embrace. His attitude is driven by the importance of his goals—to help reduce global warming, save the planet, and improve the way we all live—and by the sincerity with which he pursues them, which is rare among businesspeople.

Gary also helped open doors for us with an even more controversial corporation—the giant retailer Wal-Mart, which is seen in some quarters as a symbol of everything evil in contemporary capitalism. Gary's organic products are now stocked in Wal-Mart stores, reflecting the fact that popular understanding of the value of organic production methods has become widespread, leading to growing pressure on major retailers to support alternative sources of food. It took years for us to convince Wal-Mart to let us include that story in our movie, but Gary helped win them over.

Frankly, I think it was a smart business move on Wal-Mart's part. When you have problems with your public image, as Wal-Mart does, why not take advantage of every opportunity to show a more benign side of your operations? The scenes in *Food, Inc.* depicting Gary Hirshberg's interactions with the folks at Wal-Mart reflect a little of my original vision for the picture, as a balanced portrayal of the American food system that included corporate as well as anticorporate perspectives. There is plenty to criticize about Wal-Mart, including their role in promoting the industrialization of America's food supply as well as their dismal record on the labor front. But they are certainly not all bad, and when they do something right, they deserve to get credit for it. I still regret that most of the companies we dealt with weren't willing to open their doors the way Stonyfield and Wal-Mart did.

In the end, I think *Food, Inc.* became a movie that is mainly about power—about how a few companies have managed, in the last forty years, to take over a major segment of American society and now are doing everything they can to maintain and extend that power, including controlling our access to information about their activities. This is the thread that connects the disparate stories we tell in the picture. As one of the best early reviews of the picture (written by Andrew Goldstein in the *Los Angeles Times*) said, "The best thing about the documentary is that it does what good reporting does—it connects the dots." I loved that comment because it described exactly what we tried to achieve and suggests that, for one viewer at least, we achieved it.

———

Since we finished the movie, world events have only made the story more relevant and disturbing. In the summer of 2008, hundreds of thousands of infants in China were poisoned by milk containing the chemical melamine, which is used as a way of extending production while reducing costs and increasing profits. We in the United States have a slightly better system for policing our food supply than they have in China. But only slightly better, and the food industry is trying to lower our standards all the time. At the same time, companies try to convince us that they can do a better job of policing themselves than the government can. I don't think any company should have the right to decide what is safe and not safe without some government oversight, especially when they have billions of dollars at stake.

I have to say that the incredible crisis now sweeping the world's financial system also reminds me of the problems with the food system that we documented in *Food, Inc.* Consider these parallels. Most experts agree that the collapse of credit markets caused by the meltdown of the subprime mortgage industry was made possible by thirty to forty years of financial market deregulation—just as the cozy relationships between regulators and regulated in the food industry have, in effect, deregulated America's food supply. In both cases, the assumption is that corporations are capable of policing themselves—an assumption that both experience and human nature teach us is invalid.

Left to their own devices, banks got into derivatives and other risky, untested, and unregulated financial instruments—and almost brought down the economy with them. In the same way, corporations subject to few external controls are experimenting with our food supply, using chemical additives, hormone treatments, pesticides and fertilizers, and mechanized production methods that represent a revolutionary break with almost 10,000 years of agricultural history. In the process, as Michael Pollan has pointed out, they keep creating problems and then apply even more technological innovations as Band-Aids, rather than going back to the natural processes they discarded.

I should emphasize that I'm not anticorporate on any philosophical level. I put companies like Stonyfield and Wal-Mart into the film because I believe they can be part of the solution. Thankfully, we still have a democratic system in the United States. Ultimately, the people have the power to change the system. And because the executives who run our corporations know this, they are responsive to public opinion and will try to do the right thing if they are pressured into it.

Thus we see companies like McDonald's responding to animal-rights activists by trying to guarantee that the meat they buy is raised according to at least minimal standards of humane treatment. (We even filmed a spokesman for North Carolina pork producers who complained bitterly about this new policy on the part of McDonald's, one of many bits of film we didn't have room to include in the finished movie.)

U.S. corporations are very conscious of how they're perceived, and they worry about it a lot. That gives consumers a lot of power—if they're prepared to inform themselves and then to make smart decisions based on what they learn. As Eric Schlosser says at the end of the movie, through our actions we were able to reduce the power of the tobacco companies, which like the food companies had been extremely powerful and entrenched in government. Hopefully we can do something similar in regard to the companies that dominate our food system.

As I've said, I don't view myself as a crusader. But if *Food, Inc.* can contribute to the effort to help Americans change the way they eat for the better, I'll be pretty happy with that result.

# Another Take | FOOD SOVEREIGNTY FOR U.S. CONSUMERS

## By Food & Water Watch

As noted on page 19, Food & Water Watch is a nonprofit consumer organization that works to ensure clean water and safe food. In "Food Sovereignty for U.S. Consumers," Food & Water Watch raises an issue most Americans may never have thought about—how much control do they *really* have over the foods they eat? For more information, visit the organization's website at http://www.foodandwaterwatch.org.

Ask a shopper in any supermarket aisle if he or she supports "food sovereignty," and you'll probably get a blank stare in response. But ask most consumers some basic questions about what kind of food they want to feed their families, and they're probably closer to understanding food sovereignty than they think.

As consumers learn more about what the corporate-controlled food system is feeding them, the appeal of local, family-farmed food grows. Just think about the headlines we have seen in the last few years:

- Spinach contaminated with E. coli leads to a nationwide recall, with several deaths and hundreds of illnesses in more than twenty states linked to produce from one region of California.
- An unapproved variety of GMO rice contaminates most of the U.S. rice supply—and federal regulators respond by announcing their intention to approve the rice for human consumption, after the fact.

- The FDA intends to approve meat and milk from cloned animals.

Add these to the growing awareness of how far our food travels and concern about food-related health issues like obesity, and the increased interest in local, organic, and sustainable food starts to make sense.

So how does this get us to food sovereignty? Food sovereignty is the principle that people have the right to define their own food and agriculture system. This is a stark contrast from a system that is dominated by World Trade Organization (WTO) rules and corporate commodity traders.

Food sovereignty can benefit small producers all over the world and give consumers things that the "free trade" agenda cannot:

- The right of countries (or states) to ban a certain hormone, pesticide, or technology like genetically modified foods or irradiation, without fear of starting a trade war that will ultimately be decided by an unelected panel of lawyers at the WTO.
- The right of countries (or states) to label their food the way they want to.
- Regional food supplies—with local farmers able to get a fair price so they can stay in business and supply local markets.
- Better food! And a diversity of breeds and crop varieties, instead of the few promoted by agribusiness that can be grown with intensive techniques and stand up to a trip halfway around the planet.

## WHAT YOU CAN DO

- **Demand to know where your food is from.** Country of origin labeling (COOL) gives consumers a way to make more informed purchases and support domestic producers. COOL was included in the 2002 Farm Bill, but only the provisions for seafood were allowed to go into effect. Tell Congress that comprehensive country of origin labeling is long overdue.
- **Support your local farmers.** The options for supporting local food have never been better. You could: (1) Ask your grocery store to carry and label local food in the produce aisle. (2) Shop at your local farmers'

market. Check out the USDA's directory of farmers' markets to find a market near you and learn whether they support local growers. (Visit http://www.ams.usda.gov and click on "Wholesale and Farmers Markets.") (3) Become a member of a community-supported agriculture farm. CSAs are a great way to support a farmer directly and get great local produce, meats, and more. Find out more by visiting the USDA's web page on CSAs at http://www.nal.usda.gov/afsic/pubs/csa/csa.shtml.

- **Speak out on U.S. farm policy.** The Farm Bill is not just an arcane debate about soybean prices that takes place every few years in Washington, D.C. It sets the ground rules for our whole food policy—and it affects everybody who eats. Get updates from Food & Water Watch on the Farm Bill and other food issues by signing up at http://www .foodandwaterwatch.org/take-action/mailing-list-signup.html.

# Part II | INSIDE THE FOOD WARS

# THREE

# ORGANICS—HEALTHY FOOD, AND SO MUCH MORE*

## By Gary Hirshberg

Gary Hirshberg is chair, president, and CE-Yo of Stonyfield, the world's leading organic yogurt producer, based in Londonderry, New Hampshire.

For the past twenty-five years, Hirshberg has overseen Stonyfield's growth from its infancy as a seven-cow organic farming school in 1983 to its current $320 million in annual sales. Stonyfield has enjoyed a compounded annual growth rate of over twenty-four percent for more than eighteen years by using innovative marketing techniques that blend the company's social, environmental, and financial missions.

In 2001, Stonyfield entered into a partnership with Groupe Danone, and in 2005, Hirshberg was named managing director of Stonyfield Europe, a joint venture between the two firms with brands in Canada, Ireland, and France.

Hirshberg has won numerous awards for corporate and environmental leadership, including Global Green USA's 1999 Green Cross Millennium Award for Corporate Environmental Leadership. He was named Business Leader of the Year by *Business NH* magazine and New Hampshire's 1998 Small Business Person of the Year by the U.S. Small Business Administration. He is the chair and cofounder of O'Naturals, a natural fast-food restaurant company, and the author of *Stirring It Up: How to Make Money and Save the World* (Hyperion Books, 2008).

O n Thanksgiving 1977, during my tenure at the New Alchemy Institute—an ecological research and education center on Cape Cod—my family came to visit. My housemate, Rob, had made an organic

---

pumpkin soup for us before leaving to spend the holiday with his own family. The soup was awful. After one spoonful, I declared my innocence in this culinary crime and tossed out the evidence. "If this is organic food," my mother chimed in, "bring on the chemicals." Fortunately, the organic salad, yams, and turkey were delicious, and by the end of the dinner, Mom had come around.

Organic food has come a long way in terms of taste since that holiday meal, but organic isn't just about food. It's a much more expansive way of thinking that embraces cyclical, nonlinear resource use, where waste from one activity becomes food for another. It honors natural laws, and it abhors the mindless dispersal of toxic chemicals. Cheap substitutes don't work. That's why you can't replace organic farming with chemical farming and expect anything but depleted soil, poor crops, and unstable prices.

All of humanity ate organic food until the early part of the twentieth century, yet we've been on a chemical binge diet for about eighty years—an eye blink in planetary history—and what do we have to show for it? We've lost one-third of America's original topsoil; buried toxic waste everywhere; and polluted and depleted water systems, worsened global warming, and exacerbated ailments ranging from cancer to diabetes to obesity.

This is not airy blather touting the tofu way to happiness. I see organic as part of a philosophy of wholeness, the science of integration, the need to keep nature humming as the interdependent web of life. Organic is also a pragmatic state of mind, offering real antidotes to society's assorted ills and errors. It backs a sensible farm policy that protects not only family farmers, but also the health of all Americans—when you eat better, you are better. In fact, an organic food system could bring down health-care costs by eliminating toxic lifestyles and the unnecessary disease and illness they cause.

Organic methods of agriculture can help stabilize fuel prices and reduce our dependence on foreign oil. They can lead to true national security, which, in turn, fosters planetary security. By using less fossil fuel and chemicals, and by trapping and building carbon in the soil instead of in the atmosphere, organic farming is a crucial WME (weapon of mass enlightenment) in humanity's now-or-never fight against the air pollution that causes global warming.

In short, there's nothing "alternative" about organic.

Back in the 1970s, organic food had no such positive image. Many dismissed it as a fringy fad served cold with an eat-your-spinach sermon. How could organic taste good? Indeed, taste was the key challenge. Organic advocates couldn't popularize a cuisine simply by declaring it spiritually and ecologically superior. The world, like my mother, was not waiting for or willing to eat inedible soul food. To win acceptance, it had to be truly delectable.

But that would take a while. Many of us got involved in the organic movement for political reasons—to protest industrial agriculture. Some of us were back-to-the-land rebels with a strong passion for eating locally grown food. Others were food purists, excited by the opportunity to propagate and preserve heirloom varieties of produce and seed stocks. Still others came to the cause simply for the joy of growing our own food, talented amateurs at best who cared more about its appeal to a diner's political conscience than to his or her tastebuds.

Luckily, what began as a philosophical fondness for dishes like brown rice and seaweed eventually matured into a tasty cuisine that attracted talented chefs, notably my friend Alice Waters, who called organics "the delicious revolution."

How I found myself swept away by this revolution is a story that began in the late 1970s, when I was executive director of the New Alchemy Institute. We built a solar-heated greenhouse that used no fossil fuels, herbicides, pesticides, or chemical fertilizers. Yet it produced enough food to feed ten people three meals per day, 365 days per year. Even when the yard was covered with snow, it was toasty inside—a haven for everything from birds and bees to bananas, figs, and papayas. Tanks of water absorbed sunlight by day and radiated heat at night. Each tank raised about 100 pounds of fish per year. Their waste fertilized plants, which, in turn, provided food for the herbivorous fish. Wind systems provided electrical and mechanical power.

All this seemed to be a worthy achievement to this young idealist until 1982, when I visited my mother, then a senior buyer at Disney's Epcot Center in Orlando, Florida. There I rammed into one of those epiphanies that change your life forever.

We toured the Land Pavilion where sponsor Kraft Foods was touting its vision of future farming. In tribute to the blessings of supposedly endlessly fertile land and unlimited resources, Kraft displayed, in a building both heated and cooled by fossil fuels, rivers of chemical fertilizers, herbicides, and pesticides swooshing around the naked roots of anemic-looking plants grown hydroponically in plastic tubes. In this paean to fertility, there was not a single grain of actual soil. Natural farming is all about creating great dirt, rich with nutrients. This was a cartoon scene of chemistry gone mad. As I saw it, nothing grown the Kraft way would sustainably nourish a laboratory rat, much less soil itself.

Kraft underscored its bizarre message with "Kitchen Kabaret," a pseudo-musical featuring the four food groups and Bonnie Appetit, the show's only live character. The other "actors" were animatronic robots named Miss Cheese, Miss Ice Cream, Miss Yogurt, and the like, who sang lyrics I wish I could forget:

> Your taste buds I'll appease.
> I know how to please.
> It's known that I am too good for words.
> Oh, isn't that right, big boy?

As bad as the lyrics were, I left feeling that the singing foodstuffs were secretly humming a very different and troubling tune:

> Just buy Velveeta, please.
> So what if it's not real cheese?
> Real is what we'll say it is,
> And Mother Nature's on her knees.

Every day, 25,000 people paid to see this spectacle—more than visited my own New Alchemy Institute in a year. After viewing the Kraft-sponsored pavilion twice myself, I came away deeply disturbed.

While stewing about all this, I had a Eureka flash that eventually shaped my life. I blurted out to my mother: "I have to become Kraft."

Don't misunderstand. I was still convinced that Kraft was crazy and only sustainable practices could save the planet. But now I faced the reality that people like me were unheard voices preaching to ourselves in an uncaring world. To change anything, we needed the leverage of powerful businesses like Kraft. If we had their cash and clout, people would listen and begin to make changes. In other words, to persuade business to adopt sustainable practices, I would have to prove they were profitable.

Enter Samuel Kaymen, a self-taught yogurt-making genius from Brooklyn, New York.

Samuel, one of the most focused men I've ever known, was a brilliant former defense-industry engineer in search of a higher purpose. He had switched lifestyle gears twice, first to become a hippie and then to found the Rural Education Center, an organic farming school in Wilton, New Hampshire. Now a lively graybeard with twinkling blue eyes, Samuel had the penetrating mind of a superb engineer, but not a single business corpuscle anywhere in his body. And by 1983, his organic farming center was virtually bankrupt, on the brink of being plowed under.

At that troubling moment, hoping to help save the center, I agreed to become a trustee. I loved the concept of sustainable farming with its inherent affinity with and respect for nature, so unlike business as I knew it then. I particularly admired Samuel's ambition to help farmers stabilize their shaky finances and gain the secure lives they deserved. We were kindred spirits.

As a trustee, I first concentrated on new ideas to meet the center's $150,000 annual budget, but equally urgent was the need to help focus Samuel on priorities. He was a seat-of-the-pants manager, completely overcommitted and overwhelmed. I convinced him to eliminate unnecessary activities and put his laser-like brain to work on two fronts: fulfilling the center's mission and generating revenues. That's how we came to zero in on the yogurt business.

Samuel was already an astute yogurt maker. Certain that he had a unique product, he became an impassioned dairyman, milking his farm's cows and selling yogurt under the Stonyfield label through the Rural Education Center.

But Samuel owed money he couldn't pay back, and he was slipping further behind every day. Exhausted and financially strapped, he asked me to come aboard full-time.

I loved Samuel, a compelling eccentric on a holy mission. I loved his yogurt, the most delicious I'd ever tasted. And I loved the hills of my native New Hampshire. How could I not agree to help run his business?

The day I arrived, September 15, 1983, I opened the door to the office and found three Army surplus desks piled chest-high with teetering stacks of unpaid bills. I looked for the bank deposits to balance out the bills. There were none—Samuel had cashed every incoming check to buy feed and supplies. By 3:30 that afternoon, after opening every envelope, I'd calculated that we were $75,000 in the red. The electricity was about to be cut off. Stonyfield was a beautiful dream created by a beautiful guy. It was also virtually dead on my arrival.

So I did the only thing possible in that situation: I called my mother and borrowed money. Then I called other friends, family members, and Stonyfield trustees and managed to raise more money. Finally, I called our landlord with a vague promise of some kind of ultimate payback in lieu of rent.

My reward for all this was a demanding new life that entailed chores like helping Samuel milk the cows twice each day and make yogurt every weekend, often in subfreezing weather. But we toiled on, increasingly aware that we needed to be totally clear about why we were beating ourselves up like this—not for self-affirmation alone, but also to spread our environmental gospel and convince others to join in our cause.

We had started with vats of yogurt and a fuzzy notion of making money by persuading the world that our products could help clean up the planet and save family farmers. But we had never refined our ideas. Every world-saving (and moneymaking) business needs a stated mission—a rallying cry that focuses efforts, helps set priorities, and gives all hands a meaning and a purpose. So one night I sat down and wrote a mission statement that has barely changed since. Stonyfield is in business:

- To provide the very highest-quality, best-tasting, all-natural, and certified organic products.
- To educate consumers and producers about the value of protecting the environment and supporting family farmers and sustainable farming methods.
- To serve as a model that environmentally and socially responsible businesses can also be profitable.
- To provide a healthful, productive, and enjoyable workplace for all employees, with opportunities to gain new skills and advance personal career goals.

- To recognize our obligations to stockholders and lenders by providing an excellent return on their investment.

After twenty-five years, the mission I backed into has produced a successful company that combines profitable business with a powerful purpose. The truth is, even I had doubts back in 1983 that Stonyfield could reject conventional business wisdom, however perverse we judged it to be, and still prosper financially. Our decision was basically a gamble that a truly honest product—pure yogurt—would attract customers so in tune with our environmental goals that both our business and our ideals would flourish. The gamble paid off, but only after we made our mission clear to ourselves, our customers, our investors, and the world.

In those early days, many of us felt that our passion, in all its various forms, could bloom into businesses—if "we" grew it, "they" would buy. But that was a misjudgment that spawned lots of bankruptcies. I was committed to both the food and the politics, but I also believed that if organics were to gain traction and grow beyond our original small enclave of activists, America's supermarkets—and, more important, the way they did business—would have to be accommodated. Inevitably, this led to friction with various friends who seemed more interested in fighting culture wars than seizing new commercial opportunities. I let them go their way, and never really looked back.

On some level, I sensed that it was only a matter of time before the world came around to the diverse business possibilities of the organic way of life. But even I couldn't have foreseen back then just how huge and profitable the opportunity would become.

Today, burgeoning demand for organic food is winning over some of the biggest names in retailing, including Safeway, Kroger, Costco, Target, Publix, H-E-B, and others—even Wal-Mart, which is nothing if not savvy about where to make money. It's exciting to think of price-conscious shoppers the world over voting with their pocketbooks to save the Earth.

In Stonyfield's early days, the notion of organic going mainstream was only a dream. We faced immediate problems of supply and demand: we didn't have much of either. Sure, a few farm cooperatives were growing and selling organic products to a few buyers, but the vast majority of milk producers and consumers weren't remotely interested. We compromised, using ingredients such as milk free from artificial bovine growth hormone—which increases milk production per cow, but which many say lead to shortened lives and was ultimately prohibited by organic standards—but without imposing other restrictions on our suppliers.

The organic milk market finally began to churn in the early 1990s. The precipitating move came when Monsanto went a bridge too far in its biotech invasion of the food market: After years and millions of dollars spent on lobbying, and despite the testimony of nutrition and agriculture experts including myself, Monsanto won official approval to sell its bovine growth hormone, rBST, for use in milk cattle. Adding to the outrage, Monsanto also got the nod to leave off milk labeling the fact that milk contained this artificial ingredient. This was what made Horizon Organic Dairy and the Organic Valley Family of Farms decide to launch large-scale organic production. It was also the trigger for my own decision to make Stonyfield fully organic.

We got another break when Peter and Bunny Flint abandoned conventional dairy farming to launch their Organic Cow brand in Tunbridge, Vermont. We began working with the Flints in 1994, buying the milk we needed to convert our plain, whole-milk yogurt from natural to organic, first in a one-quart container and then in six-ounce cups. It was a gamble from the first spoonful because our yogurt was already pricier than every other conventional brand and would have to increase cost even further. Having committed ourselves to keeping a high floor price for organic milk—conversion from conventional dairying to organic means a financial commitment by farmers, who struggle to make ends meet during the required three-year conversion period—we had to price Stonyfield yogurt high enough to cover the expense of paying the Flints and our other suppliers a hefty premium over conventionally produced milk. And when we added fruit and sugar to the mix, our production costs rose still higher. Like milk, organically grown sugar and fruit were priced twice as high as conventional ingredients.

Why this higher price? Organic regulations require that farmers use cultural, biological, and physical methods in protecting crops, and these often take longer to yield results than indiscriminate use of chemicals, fertilizers, and the like. Careful management of the soil and the avoidance of antibiotics to treat sick animals also take patience, although in the case of organic dairy cows, they're likely to be healthier in the first place. The yield per organic acre or per animal often goes down, at least during the first years after conversion. Still, organic farmers say they can ultimately exceed the yields of conventional rivals through smarter soil management.

Fortunately, our customers did not flinch. They kept paying our price for six ounces of our organic yogurt, even though they could buy eight ounces of a conventional brand for the same price. Of course, we were, in a sense, stealing from Peter to pay Paul, because our own nonorganic natural products (then ninety percent of our business) were subsidizing the switch to organic.

By 2001, our volume was rising and our costs were dropping, with organic sugar on its way to price parity with conventionally grown sweetener in 2003 as larger organic operations entered the supply pipeline. The unbeatable combination of growth and savings kept us alive and steered us to prosperity. At long last, in the fall of 2007, we converted the last fifteen percent of our products to organic. Being "natural" is a good thing, but despite all the talk about natural foods these days, we still lack tough standards defining exactly what the term means for consumers. Moreover, my big concern is that farmers raising natural food crops have no guarantee they'll be fairly compensated, regardless of how much time and money they spend protecting the Earth. The demand for their products remains a very small fraction of the consumer demand for conventional products, so with a smaller market comes a smaller payback.

The organic food business is the only industry I know that actively seeks increased government regulation. It's our way of remedying public confusion about the meaning of organic. In 1979, the need for clear standards led my partner Samuel to form the Natural Organic Farming Association (now the Northeast Organic Farming Association), the first of many private and state agencies that eventually certified farms and foods as truly organic. In turn, the need to clarify local disparities led Congress to pass the federal law ordering the U.S. Department of Agriculture (USDA) to create a single set of nationwide standards.

Completed in 2002, those criteria strictly define three levels of organic labeling. The term "100% Organic" is fairly self-explanatory and refers to foods and fibers that are indeed produced organically at every step, from farm field to store shelf. The second level, simply "Organic," requires that at least ninety-five percent of a product's ingredients be organic, with the remaining five percent strictly limited to ingredients on USDA's National List of Allowable and Prohibited Materials. To certify our vanilla yogurt as organic, for example, inspectors must inspect and certify multiple supply sources—the farms producing our organic milk, sugar, and vanilla beans, and the company turning the beans into our vanilla extract—along with our own Yogurt Works. The third category, "Made with Organic," means that at least seventy percent of a product's ingredients are organic.

The upshot is that the word "organic" is more credible than ever. But I still wish consumers were given more exacting data about the nutritional content and benefits of all their foods. Knowing more about how the food we buy is produced might shake otherwise heedless consumers and get them asking questions about many conventional food-growing methods. Maybe they would want to know just what kind of food and antibiotics were pumped into the animals that end up as beef steaks, pork chops, and chicken tenders on our tables. Or perhaps they would be interested to learn which pesticide was sprayed on their fruits and vegetables, along with its potential health effects. Cheaper at the checkout might not seem like such a bargain if the longer-term consequences were more readily apparent. As it is, too many consumers are lulled into complacency by a lack of meaningful information. In our case, we provide complete nutritional information on our packaging and on our website. We can tell you what goes into every cup, and we're happy to report there are no numbers or strange chemistry formulations to be found.

Organic food is a thriving, $18 billion-a-year business in the United States. Given the market's steady growth, mainline food companies have hastened to acquire or create their own organic brands, among them Heinz Organic Tomato Ketchup and Kellogg's Organic Raisin Bran, Rice Krispies, and Mini-Wheats cereals.

With the likes of Wal-Mart and Target putting pressure on organic suppliers to cut prices, and with companies like Kellogg's and Heinz joining the competition, some of my organic colleagues worry that we will lose our ability to charge a premium and be forced to start undermining farmers' prices, along with quality and standards. But I welcome the trend. There is no denying that the turn toward organic products is a big plus for America's health and our planet's environment. If the good stuff becomes commoditized, and if entrepreneurs continually raise the bar by adding in extra nutrition, there will be less bad stuff on the market and less strain on our environment. And isn't that our ultimate goal? If we are really committed to saving the world, we'll stop questioning motives and start applauding every big company that makes the effort to change course and embrace organic methods.

There are dangers to the trend, however. Various large corporations are buying their way into the organic market by acquiring established players—and sadly, many of them are weakening and diluting these brands. Anyone with enough money can buy a company, but it takes a real commitment to the core principles of organic farming and business methods to nurture it and make it work. And while it is not possible to dilute organic standards due to stringent regulation and third-party audits, some companies with social and environmental missions have seen those missions falling by the wayside after being swallowed by larger corporations and as the bottom line becomes a top priority.

Not so with Stonyfield. We have a unique partnership with Groupe Danone, the France-based consumer products company well known for its biscuits, its Evian brand bottled water, and its Dannon yogurt.

Between 2001 and 2003, Groupe Danone bought about eighty-five percent of our company shares. Typically, when big conglomerates buy smaller, organic, or natural food companies (Kraft and Boca Burger, Nestle and Power Bar, Unilever and Ben & Jerry's) the smaller company undergoes management changes. But Groupe Danone made no changes to our management team, and no changes to our company's environmental and social missions.

I remain Stonyfield's full-time chair, president and "CE-Yo." Through our Profits for the Planet program, we still give ten percent of our profits to environmental causes. Our milk still comes from New England and Midwest dairy farmers through the CROPP cooperative. We still use the very best environmental practices we can find. And we're still as committed as ever to increasing the number of organic family farms in the world.

Groupe Danone has brought a lot of new knowledge and talent to our company. Our partner is helping us manage our rapid growth by sharing its expertise in food research, production, operations, logistics, and distribution. On the other hand, the partnership allows Groupe Danone to participate in the organic and natural dairy business, which is growing at a rate rarely seen in the food industry.

In Groupe Danone, we've found a partner that provides cost-saving efficiencies and growth opportunities, while allowing us to manage ourselves autonomously and remain true to our mission.

I can't help but think back to my "aha" moment in the Kraft Food Pavilion. Ever since that moment, when I realized that businesses must adopt sustainable practices—for the sake of our health through organic foods and for the Earth's sake through more environmentally friendly practices—the challenge of proving that sustainability pays—and hugely—has driven my career. Somewhat to my own surprise, I have succeeded. After years of work and many experiments, I have discovered that sustainable practices not only make money but are invariably more profitable than conventional business methods.

Once corporations such as Kraft realize that businesses can derive big profits from cleaning up the planet and operating in green, sustainable ways, the battle will be won. Business is the most powerful force on Earth. Unlike governments, which are usually bound by consensus and convention, business can lead. Unlike churches, community groups, and nonprofits, business has money to back up its ideas. It can act quickly, get rules changed, and overcome entrenched interests. In one of those ironic twists that make life so interesting, the same boundless thirst for profit that got the planet into trouble can also get us out of it.

Take climate change. From 1995 to 2005, Stonyfield had great success in increasing our efficiency and slashing our carbon dioxide emissions per ton of yogurt we produced. Carbon dioxide emissions contribute to global warming, the gravest environmental threat of our time. Our pollution-cutting efforts were equivalent to taking 4,500 cars off the road. Moreover, we saved more than $1.6 million in the process—good reason for all companies to do the same.

In 1997, we were the first manufacturer in the United States to mitigate the $CO_2$ emissions from our facility by investing in carbon offsets—such as reforestation projects and the construction of wind generators.

Just as we've profited by reducing $CO_2$ emissions, so we've further reduced Stonyfield's environmental footprint by rethinking waste management. The human tradition, alas, has been to "dilute" garbage by dumping it in the nearest moving water, which takes it to a mythical place called "away," meaning simply out of sight. The new notion of recycling is more enlightened, but only slightly. Had the waste been avoided or reduced in the first place, it would not have to be recycled, with attendant costs for removing and reusing it.

Conversely, we can't ever match nature's ability to waste nothing by reusing everything in closed, self-sustaining systems. We still generate waste, such as tons of yogurt that accumulates from the quality-control process. The strategy that works for us is to sharply reduce certain materials before they begin generating waste. For example, after completing an environmental assessment of our packaging, we eliminated our plastic lid and inner seal in favor of a foil seal, dramatically reducing the amount of packaging material and the energy and water that went into making it. That change alone saved more than $1 million a year.

Whatever the variations from one business to the next, twenty-five years of hard-won success at Stonyfield tell me that sustainability is the correct grail for business leaders to pursue in the early twenty-first century. Those who do will reap unimagined rewards for themselves and the planet. This is no longer a hypothesis; it's a surefire plan for solid transformation.

The fact is that eco-solutions (to me, that prefix signifies both ecological and economic) like organics, waste reduction, and greenhouse gas (GHG) reductions present the biggest business opportunities in the history of humankind.

Organic stands for many things—a philosophy of wholeness, the science of integration, a rallying cry for keeping nature humming as the interdependent web of life. Organic is also highly pragmatic—a real solution to society's ills. It's a sensible farm policy and helps mitigate health-care woes—you eat better, you are better. It can help stabilize fuel prices and reduce our dependence on foreign oil. It can lead to true national security. By using less fossil fuels and chemicals, and by trapping and building carbon in the soil instead of the atmosphere, organic is a profound weapon in the fight against air pollution and global warming, a matter of planetary security.

In short, organics is about more than food. It's about survival.

| *Another* | THE DIRTY SIX |
| *Take* | *THE WORST ANIMAL PRACTICES* |
| | *IN AGRIBUSINESS* |

## By the Humane Society of the United States

---

As the largest animal protection organization in the nation, The Humane Society of the United States (HSUS) takes a leadership role on farm animal advocacy issues. In "The Dirty Six," HSUS explains the six most abusive practices employed by agribusiness in the raising of animals for human consumption. For much more information on the treatment of farm animals and HSUS's role in advocating on their behalf, visit the organization's website at http://www.hsus.org/farm.

---

In just one hour in the United States, more than one million land animals are killed for food. Before their slaughter, most of these farm animals—nearly 10 billion each year—endure lives of abuse with virtually no legal protection at all. Considering this staggering figure, the mistreatment of farm animals is among the gravest animal welfare problems in the nation. Instead of being recognized as the social, intelligent individuals they are, chickens, pigs, cows, turkeys, and other animals are treated as mere meat-, egg-, and milk-production units and denied expression of many natural behaviors. And six standard agribusiness practices are the most egregious of all.

1. **Battery Cages**

    In the United States, approximately ninety-five percent of egg-laying hens are intensively confined in tiny, barren "battery cages"—wire enclosures stacked several tiers high, extending down long rows inside windowless warehouses. The cages offer less space per hen than the area of a single sheet of paper. Severely restricted inside the barren cages, the birds are unable to engage in nearly any of their natural habits, including nesting, perching, walking, dust bathing, foraging, or even spreading their wings.

While many countries are banning the abusive battery cage system, U.S. egg producers still overcrowd about 300 million hens in these cruel enclosures.

## 2. Fast Growth of Birds

More than nine out of ten land animals killed for human consumption in the United States are chickens raised for meat—called "broilers" by the industry. About nine billion of these birds are slaughtered every year. According to poultry welfare expert Ian Duncan, PhD, "Without a doubt, the biggest welfare problems for meat birds are those associated with fast growth." The chicken industry's selective breeding for fast-growing animals and use of growth-promoting antibiotics have produced birds whose bodies struggle to function and are on the verge of structural collapse. To put this growth rate into perspective, the University of Arkansas reports that if humans grew as fast as today's chickens, we'd weigh 349 pounds by our second birthday.

Consequently, ninety percent of chickens raised for meat have detectable leg problems and structural deformities, and more than twenty-five percent suffer from chronic pain as a result of bone disease.

## 3. Forced Feeding for Foie Gras

French for "fatty liver," the delicacy known as paté de foie gras is produced from the grossly enlarged liver of a duck or goose. Two to three times daily for several weeks, birds raised for foie gras are force-fed enormous quantities of food through a long pipe thrust down their throats into their stomachs. This deliberate overfeeding causes the birds' livers to swell as much as ten times their normal size, seriously impairing liver function, expanding their abdomens, and making movements as simple as standing or walking difficult and painful. Several European countries have banned the force-feeding of birds for foie gras, and the state of California is phasing it out. The

United Nations Food and Agriculture Organization (FAO) states that the "production of fatty liver for foie gras . . . raises serious animal welfare issues and it is not a practice that is condoned by FAO."

4.  **Gestation Crates and Veal Crates**

    During their four-month pregnancies, sixty to seventy percent of female pigs, or sows, in the United States are kept in desolate "gestation crates"—individual metal stalls so small and narrow the animals can't even turn around or move more than a step forward or backward. The state of Florida and the European Union (EU) have already begun phasing out the use of gestation crates because of their inherent cruelty, yet these inhumane enclosures are still the normal agribusiness practice of most U.S. pork producers.

    Similarly, most calves raised for veal are confined in restrictive crates—generally chained by the neck—that also prohibit them from turning around. The frustration of natural behaviors takes an enormous mental and physical toll on the animals. As with gestation crates for pregnant pigs and battery cages for egg-laying hens, veal crates are widely known for their abusive nature and are being phased out in the EU but are still in use in the United States.

5.  **Long-Distance Transport**

    Billions of farm animals endure the rigors of transport each year in the United States, with millions of pigs, cows, and "spent" egg-laying hens traveling across the country. Overcrowded onto trucks that do not provide any protection from temperature extremes, animals travel long distances without food, water, or rest. The conditions are so stressful that in-transit death is considered common.

6.  **Electric Stunning of Birds**

    At the slaughter plant, birds are moved off trucks, dumped from transport crates onto conveyors, and hung upside down by their legs in shackles. Their heads pass through electrified baths of water, intended to immobilize them before their throats are slit. From beginning to end, the entire process is filled with pain and suffering.

    Federal regulations do not require that chickens, turkeys, and other birds be rendered insensible to pain before they are slaughtered. The shackling of the birds causes incredible pain in the animals, many of whom already suffer leg disorders or broken bones, and electric stunning has been found to be ineffective in consistently inducing unconsciousness.

## YOU CAN HELP

1. Lend your voice to help protect farm animals from suffering. The HSUS works on a number of issues, including encouraging producers to phase out the Dirty Six. Join us today.
2. Ask your local grocers not to buy animal products produced through these means. For example, encourage your neighborhood market to sell eggs only from uncaged birds.
3. Don't support the cruelties endured by farm animals—follow the Three Rs:
   - Refine your diet by eliminating the most abusive animal products
   - Reduce your consumption of animal products
   - Replace the animal products in your diet with vegetarian options

Need help getting started? Visit www.HumaneEating.org for more information on how you can help farm animals when you eat, including delicious recipes, tips on incorporating more animal-free meals into your diet, shopping list suggestions, and much more. And for more information on the lives of farm animals, visit www.FarmAnimal Welfare.org.

**FOUR**

# FOOD, SCIENCE, AND THE CHALLENGE OF WORLD HUNGER—WHO WILL CONTROL THE FUTURE?

## By Peter Pringle

---

Peter Pringle is the author of several books on science, business, and politics, including *Food, Inc., Mendel to Monsanto—The Promises and Perils of the Biotech Harvest*, and *Cornered, Big Tobacco at the Bar of Justice*. His latest book, *The Murder of Nikolai Vavilov*, is about Stalin's persecution of the great twentieth-century Russian plant hunter. His novel *Day of the Dandelion* is a thriller about agribusiness and biotech patents. For thirty years Pringle was a foreign correspondent for *The Sunday Times* of London, *The Observer*, and *The Independent* in Europe, the Middle East, Africa, and the former Soviet Union. He has also written for *The New York Times*, *The Washington Post*, *The Atlantic*, *The New Republic*, and *The Nation*. He is a graduate of Oxford University and a fellow of the Linnean Society of London. He lives in New York City.

---

A decade ago, two European biotech plant researchers found a way to insert a daffodil gene into a rice plant. When the plant matured, its pearly white grains turned a translucent yellow. Presto! The world had a biotech miracle food—golden rice.

The new rice grains contained beta-carotene, the substance that puts the color yellow in corn and gives carrots their traditional orange hue. In humans, it becomes the essential nutrient known as vitamin A. And lack of vitamin A causes death and blindness in millions of undernourished people in Asia and Africa where rice is a staple food. In theory, the golden rice with beta-carotene could save millions of lives. Better still, since the rice had been created with public funds from governments and foundations rather than investment by agribusiness, its seeds could theoretically be given away free to small farmers in developing countries.

There was a catch, of course. Golden rice, as its critics would later charge, was "fool's gold." It could not be distributed free to poor farmers because biotech companies owned more than forty patents on the various methods and lab tools used to create the new variety. The invention had been funded by public money, but this genetically modified (GM) plant was effectively owned by agribusiness. Anyone planting golden rice seeds would have to pay royalties on the patents.

To head off a public relations uproar, the patent-owning companies waived the royalties for poor farmers in developing countries. But anti-GM critics raised safety and environmental concerns about all GM crops; the new rice became bogged down in field trials and entangled in a web of government regulations. The botanical sensation of golden rice was soon all but forgotten.

Then came the world food crisis of 2008. Across the globe, food prices suddenly shot up, the result, in large part, of international trade policies, market speculation, and a rush to plant crops for biofuels. The crisis became so bad that the World Bank and the U.N. Food and Agriculture Organization forecast that another one hundred million people would soon be added to the 850 million who went to bed hungry every night.

As food prices surged, global warming threatened to make the crisis even worse. Farmers in developed countries responded by planting more crops and producing a record grain harvest in 2008–2009. But Asian, and especially African, farmers who lacked seeds and fertilizers could not respond so easily to the emergency. As forecasts of food shortages grew more alarming, governments in Asia and Africa came under pressure to ease their opposition to genetically modified seeds. Golden rice and other "miracle crops" were suddenly back in fashion as possible ways to help feed the world's hungry.

By 2009, there was an entirely new golden rice product with an entirely new public patron: The Bill and Melinda Gates Foundation. Better known for its work on health issues, the Gates Foundation had now entered agriculture. Hoping to spur a new Green Revolution to help increase food supplies in Asia and Africa, the Gates Foundation had supported development of a golden rice variety in the Philippines known as 3-in-1. In addition to higher levels of beta-carotene in the rice grains, this new plant also had genes built in to resist infections of bacterial leaf blight and a local rice virus known as tungro.

Yet despite the apparent potential of 3-in-1 to relieve hunger and malnutrition, anti-GM activists remained opposed to the adoption of golden rice, insisting, as they had always done, that all farmers and their governments should refuse to grow "frankenfoods" of any kind. The Gates golden rice, they argued, was a Trojan horse. Agribusiness was using the food crisis to gain public acceptance of GM crops by claiming the miracle plants could end world hunger, they insisted. Once opposition to GM crops had been overcome, they feared that the big companies would gain unlimited control through patents over the genetic basis of the world's food supplies.

The story of golden rice—its invention a decade ago, its near-oblivion, its revival by the Gates Foundation, and, above all, the intense controversy in which it remains mired—is again part of the fierce debate over biotechnology. Today, in the overfed world, people can choose not to grow GM crops or to eat GM foods. They don't need them. For the developing world the question has always been whether and how biotechnology could provide, or make a contribution to, a technological solution to the terrible problems of hunger in Asia and Africa. And could these crops be grown without surrendering a nation's future food security to a handful of global biotech corporations?

## HUMANS REMAKE THE PRODUCTS OF NATURE

To answer these questions we need to agree on what we mean by *biotechnology*. GM plants with alien genes inserted into their genomes are just one part of the broader phenomenon of biotechnology, which includes all the tools in the molecular biologist's lab box for modifying living things in ways intended to be beneficial to humans.

Plant breeders improve varieties in the traditional way farmers have done over millennia by crossing a good-looking, sturdy, disease-resistant plant with another that has a different trait for, say, high yields. Biotechnology pursues similar goals by making full use of our growing knowledge of crop genomes, waking up sleeping genes as well as silencing those not needed. The new breeding methods biotechnologists have developed include *tissue culture, anther culture*, and *gene marker selection.*

Tissue culture is growing whole plants from single cells, or plant cultures, in the laboratory instead of from seed in the soil. Under properly sterile conditions, this technique can produce plantlets without some of the diseases of their parents. Tissue culture is being used, for example, to increase the speed of production of new varieties of African staple crops such as cassava, sweet potato, and banana.

Anther culture is a special form of tissue culture. The anthers contain the pollen, or the male cell for reproduction. Breeders select pollen to produce new varieties that breed "true"—keep newly developed character traits. Using anther culture, African breeders have crossed high-yielding Asian rice (*Oryza sativa*) with African rice (*Oryza glaberrima*) that includes such desirable traits as early maturity, ability to survive among weeds, and tolerance to drought. The resulting combination is a much higher-yielding plant able to withstand Africa's different climates and soils. Dozens of varieties of this new rice combination are now being grown in West Africa, the continent's premier rice-growing region. Yields have more than doubled.

Finally, gene marker selection means linking a molecular DNA "tag" to the genes on the chromosome that govern a special trait you want to breed into a new variety. This method has proved especially useful for improving traits like root depth and disease resistance. Once the genes governing a trait have been linked to a DNA "tag," the desired trait can be moved by traditional cross-breeding to the new variety—and the "tag" shows the new trait is present. This method can cut the breeding time for a new variety in half—from an average of ten to twelve generations down to six. Marker selection is particularly useful in breeding varieties with traits that involve several genes, such as those that make leaves stay green longer and thus delay a plant drying up in a drought.

Armed with these new techniques, modern biotechnology has taken crop breeding well beyond the present reach of small farmers and local co-operatives. And therein lies a large part of the risk that has many people worried. The difference between traditional and modern techniques for modifying the genetics of living things lies not merely in the complexity of the science involved but in the economics of social and corporate control implied by the new methods.

Consider rice again. Africans are the largest importers of rice in the world today (forty-five percent of consumption)—and Africa is the fastest-growing rice market. There is a dire need to raise farm productivity. Let's say farmers produce a new rice variety that has been inbred to eliminate unwanted traits. And they have used traditional plant breeding. The new variety then has to be tested in fields free of weeds because rice is especially prone to being contaminated by its wild, weedy relatives. Before the seed is ready for sale, the breeder must improve the variety for a host of other desirable traits such as disease and insect resistance, days to flowering, and ability to withstand overcrowding so that the farmer can increase the number of plants per acre. And to sell the seed, the seed merchant must assure the quality and cleanliness of the seeds or the customers will lose confidence in them.

This process of breeding high-quality seed is not something that the average farmer does, or can afford to do. In the West, it was done traditionally by public government–run agriculture services. Farmers used to exchange their expertise about seeds, and public-sector research and breeding programs would provide improved genetic lines to farmers free of charge, treating them as "global public goods." Big business is not really interested in these seeds because they can't protect the new varieties. For example, wheat and rice are self-fertilized, meaning that the plants themselves produce the same variety of seed the following year. Any breeder can take the seed and put it on the market under a new name without fear of effective legal challenge.

High-yielding hybrid seeds are a different story. Hybrids are created from two inbred parents which are then crossed and, if properly designed, may give hugely increased yield. But if seed from the hybrid is planted, it won't produce the same high yield the next season. The only way to keep the high yield is to cross inbred varieties each year. It's an expensive operation, but the companies that can perform it can make a good profit since the farmers have to buy their seed each year if they wish to enjoy the yield advantages.

At this point, agribusiness becomes very interested. In fact, when corporate scientists began to develop hybrids—especially in maize, or corn as we know it in America—the new varieties practically ended the long-standing tradition in the developed world of having public-sector organizations develop better crops for farmers on a not-for-profit basis. How might hybrids affect the African rice market? The Chinese were the first country to produce hybrid rice on a large scale. It gave high yields, but the rice was of poor quality. Today U.S., Chinese, Indian, and European agribusiness companies are increasing their research in this sector. Before long they may be in a position to deliver hybrid rice to Africa. And not long after that they might develop a whole new line of genetically modified rice with special traits inserted to resist pests, droughts, or floods, all new varieties covered by a host of patents, just like golden rice.

## A CLIMATE OF FEAR

As science, business, agriculture, and social needs converge around these rapidly changing conditions, there is ample space for uncertainty, anxiety, doubt, and suspicion.

In particular, opponents of genetic modification worry that corporations eager to seize control of global agriculture are using the current food crisis as an excuse to foist patented GM seeds on farmers around the world. Evidence is to be found, they say, in the sudden rush to produce second-generation or so-called climate-ready GM plants designed to withstand new environmental stresses of climate change. (The first generation of GM crops dealt with biotic stress, that is, pests and weeds. The second generation focuses on abiotic stress, the result of drought, floods, heat, cold, salinity, and acidity.)

Among the new botanical wonders are plants that take up water more easily through their roots and use water and nitrogen from the soil more efficiently. (Nitrogen fertilizers are costly and add to soil acidity and pollute water tables. Half of the nitrogen that farmers apply to the soil is never taken up and used by the plant.) Now researchers have isolated genes that govern root growth so that they can design plants that will build better root systems.

There are also completely new plants altered to improve the nutritional qualities of crops. A new GM soybean increases omega-3 acids in the blood. Found naturally in salmon, trout, and fresh tuna, these fatty acids are known to protect against heart disease and diabetes and to help the growth of brain cells in the young. According to the new research, people may be able to eat soy instead of fish in order to obtain their essential omega-3s. It would be a double benefit—a healthier diet and less stress on fish stocks in our increasingly overfished oceans.

At the same time, agribusiness companies paint their own rosy picture of the overall success of GM crops. In the last five years, acreage planted with transgenic varieties has tripled in developing countries and doubled worldwide. If the trend continues, GM crops will soon cover up to ten percent of the world's farmland. In 2007, the global revenues from these patented seeds exceeded $22 billion, a ten percent increase over the previous year.

But there is something missing from this picture. Only four big crops—soy, maize, cotton, and canola—have been genetically altered with any real market impact. Where is the biodiversity essential for sustainable agriculture on small farms in Asia and Africa? the greens ask. Fewer than a dozen crops are used as staples worldwide, only 150 crop species are grown commercially, but another 7,000 play key roles in poor people's diets. What will happen to all those species in a world where economic incentives increasingly favor the planting of just a handful of high-profit GM crops?

In the renewed debate over GM plants, old concerns have resurfaced. Could the modified genes flow to other plants and destroy our botanical heritage? Are they safe to eat, or will they create new allergies in humans? Do GM crops modified for pests and weeds really reduce the use of chemicals, as claimed? If a country in the developing world "goes GM," will it be able to sell its farm products to markets closed to GM products, like Europe? Does farming GM crops lead inevitably to large-scale, industrial agriculture, squeezing out the small farms that support tens of millions of people in the developing world?

Today, the war of words over GM foods continues in the popular media. Yet most recent studies suggest that the fears of some GM opponents may be overblown. Increasingly, GM critics are driven by a general distrust of large corporations and a growing disenchantment with globalization. Consumers still lack a reasoned, accessible debate. The Vatican has declared that genetic manipulation altering DNA is a mortal sin. Some extreme leaders of the anti-GM movement, like Great Britain's Prince Charles, are simply antiscience. They condemn biotechnology as unnecessary, antinature, and ungodly. And certain countries, mostly in Europe, still refuse to grow GM crops or import GM foods (although some of these barriers are coming down).

The few reasoned economic and sociopolitical critiques are largely overlooked by opponents and proponents. The debate we really need now should focus on economic, environmental, and social strategy and the science and technology needed to achieve that goal.

## HUNGER IN THE DEVELOPING WORLD—THE BIOTECH FACTOR

In the developing world, the issues about the use of biotechnology have always been more complex. There, farm productivity and food supplies represent a vital economic and political challenge. Food crises—like the recent price hikes—point to the urgent need to develop sustainable agriculture. Let's look at sub-Saharan Africa, for example. Roughly two-thirds of the population live in rural areas and are dependent on agriculture for their livelihood. One-third are undernourished. Rainfall is often too little, or too much, most land is not irrigated, labor is scarce and mechanization is mostly absent, the cost of fertilizer is high, and access to markets is erratic and inefficient.

As a result of these and other factors, food production in Africa has not kept pace with population increase. Cereal yields are one-quarter of the world average. In the four decades between 1960 and 2000, cereal production more than doubled, but ninety percent of the increase was due to expansion of the land under cultivation. In the same period, the population more than tripled to 700 million.

In the face of this enormous challenge, reducing malnutrition will require a combination of changes: access to good-quality, affordable seeds that will grow in local environments; ready credit for farmers to buy fertilizers; access to water; stable political environments; and efficient means of distribution and entry to markets. Access to land is moving up the list as such countries as China, India, and Japan buy tracts of foreign farmland—not just in African countries like Uganda and Sudan, but also in Brazil, Burma, Cambodia, Thailand, Vietnam, and Pakistan—for their own offshore food production.

Biotechnology is now playing a positive role in improving the food production efforts, but the first two decades of the biotech revolution offered little benefit to the poor nations of sub-Saharan Africa. Even some biotech creations that might have been useful to African farmers remained on the laboratory shelf in America and Europe. For example, in Africa, a parasitic weed of the genus *Striga* inserts a sort of underground hypodermic needle into the roots of corn and sorghum, sucking off water and nutrients. A hundred million people on small farms in Africa lose some or all of their crops to *Striga*. A U.S. biotech company developed a biotech defense for this scourge, but it never reached Africa because Africans had neither the scientific infrastructure to develop it nor the funds to pay for it. Only recently have the Rockefeller and Gates foundations provided funds to develop the *Striga* defense and make it available to African farmers.

In similar fashion, European researchers tinkering with banana plants altered the banana genome to produce big floppy leaves that could resist a devastating airborne fungus called Black Sigatoka, but the remedy sat on a lab bench in Belgium for almost a decade.

Genetically modified seeds are not a crucial variable in addressing the African hunger crisis. Africans have always acknowledged that there is no single biotech fix or genetically modified miracle plant that will solve Africa's food shortages. But African agronomists concede that for some problems a GM crop variety may offer the best long-term solution.

Consider rice again. In Asia, where there is more water available than in Africa (at least for the moment), most of the rice is grown in paddies. The reason is not because rice prefers to grow in standing water but because water controls weeds and most weed species cannot grow in standing water. However, in most places in Africa—the exceptions include irrigated desert in Sudan and flatlands in Mozambique—farmers can't flood their fields because there isn't enough rain. And even if they could, they wouldn't want to, because standing water is a breeding ground for malaria mosquitoes. So weeding is done by manual labor—generally women and children who do this work at the expense of their education and health—and where manual labor is not available, as is often the case in AIDS-ravaged communities, a reasonable alternative might be a GM crop resistant to herbicides.

Recognizing the potential value of biotechnology, in all its manifestations, African governments in countries such as Kenya, Uganda, Zimbabwe, Egypt, and South Africa are expanding their agricultural research funding. Although agricultural investment accounts for only four percent of national spending in Africa, heads of state recently agreed to increase this to ten percent. And biotech is a central focus of this new spending.

As I write, Kenya, after years of debate, seems set to pass a biosafety law, the first step to the introduction of GM crops. Ugandan researchers have recently found a gene from sweet pepper that may be effective in fighting banana bacterial wilt disease. More than a score of food crops, including corn, bananas, wheat, cassava, and potatoes, are being genetically modified to make them resistant to insects, plagues, and drought. In Egypt, Algeria, and Tunisia, scientists are working on a drought-resistant and salt-tolerant strain of barley. In South Africa, where farmers already grow GM maize, they are designing a GM potato resistant to the voracious potato tuber moth.

The outside world is also paying more attention to the challenge of African food security—after many years of neglect. Although overall international assistance to Africa has risen, most of it has gone into emergency humanitarian aid. In 2007, more than thirty percent of overall aid assistance went to emergencies, including direct food aid, and only four percent to agricultural development, a figure that was twenty-six percent in the late 1980s.

Now public money is being poured into African agriculture. In 2008, the Gates Foundation joined the Rockefeller Foundation in establishing the Alliance for a Green Revolution in Africa (AGRA). By early 2009, AGRA will have invested more than $500 million in seeds, soil improvement, irrigation, improved access to markets, and extension services to farmers. The plan includes public-private partnerships that can develop and market new crop varieties. Together, the foundations are committed to providing hundreds of millions of dollars to fund all methods of biotech, from traditional crossbreeding, breeding to get rid of unwanted traits, and eventually genetic modification.

The plan is also to train African technicians to develop their own new varieties of crops by facilitating the transfer of royalty-free biotech material and knowledge. Gates is funding not only GM golden rice but also GM cassava, sorghum, and bananas—a Ugandan banana designed and made by Ugandans in partnership with Australian scientists.

However, the influx of funds is not without controversy. Gates's arrival in the southern hemisphere is being greeted scornfully by environmental groups and other nongovernmental organizations who see the foundation as yet another stalking horse for agribusiness. They are concerned about the eventual multinational control of agriculture through seeds and chemicals favoring the wealthier farmers, leaving poorer producers in debt. Ultimately, they fear the small farms will be replaced by large industrial farms, throwing many out of work.

"What is at stake here is the very future of the continent's agricultural practices—what is grown, how its grown, who grows it, who processes it, who sells it and . . . how much the African consumer will pay," the Kenyan political columnist Mukoma Wa Ngugi wrote in a critique of the Gates Foundation's work. Still, there are those pushing hard for Africans to adopt the new technologies. Kenyan-born Calestous Juma, who is professor of international development at Harvard's Kennedy School, argues vigorously that biotechnology "in the fullest sense of the word" (that is, including tissue culture, marker selection, and transgenic crops) is critical to Africa's development. Others have argued that there is an ethical obligation to search for new crops by all means available, including altering genomes. It's as important to assess the risks of using GM crops and provide rigorous safety and environmental regulations as it is to consider the risks of *not* adopting them.

## PUTTING TECHNOLOGICAL TOOLS IN THE HANDS OF THOSE WHO NEED THEM

Which brings us back to golden rice. The problem for the developing nations is that the tools needed to support this kind of innovation are not freely available. To use the current array of technologies to create a GM rice with, say, extra nutritious content, better yield, pest resistance, and improved nitrogen efficiency that grows well in West Africa would mean stepping on some very big corporate toes. Almost every step in the process of genetic engineering has been patented by agribusiness. Thus, creating new crops is like trying to solve a jigsaw puzzle with Dow or Monsanto charging you for every piece, if indeed the pieces are even available. There's not much incentive to finish the puzzle.

Recently, several public-sector colleges and agricultural and plant institutes in the United States interested in producing GM varieties for Africa and Asia have joined together to widen the scope of the licenses they obtain from agribusiness. With funds from the Rockefeller and McKnight foundations, they are using patented biotech tools—genes as well as lab techniques—free of royalties when their work is designed specifically for poor countries.

But the best solutions may yet be around the corner. One of the most radical is the formation of open-source movements to pool patents, provide improved databases and search capacities, and develop "workarounds"—new, free ways of practicing the scientific techniques that are currently patented.

The American/Australian molecular biologist Richard Jefferson is the leader of this "open innovation" movement. He has called for the creation of a "protected technology commons" where scientists can freely collaborate, innovate, share improvements, and exchange information with confidence that the work product can be used by all in the public sector to create real innovations. Jefferson's plan is to give scientists an opportunity and the incentive to contribute biotech inventions, especially lab tools, to a publicly accessible commons that is protected from private appropriation. Contributors would assign, provide, or license their inventions to the commons, which may use a patent, trademark, copyright, or other legally binding agreement to ensure that the invention is protected.

Users will be required to sign an open source license agreeing to put back into the protected commons any improvements they add, thus making the free technology self-perpetuating. Jefferson hopes that, within a decade, "biologists, businesspeople and citizens will be empowered to develop remarkable innovations that will benefit human well-being."

It's the same idea as the open-source software movement that has spawned such powerful innovations as the Linux operating system for personal computers. But it has a problem, critics point out. Biotechnology is unlikely to be carried out in the garage, and the research is often expensive. These same critics argue that biotechnologists will want some return for their relatively time-consuming and costly work. But Jefferson's answer is to mobilize the public sector—which has already been paid for its biotech contribution—to rebuild a toolkit that all parties, public and private, can use safely.

Jefferson is determined to find ways of meeting his challenges. "In solving the major crises in public health and agriculture, science has not been used sufficiently as an instrument for economic and social development," he says. "The explosion of patenting and pace of discoveries and investment in biological sciences has continually marginalized those most in need. This cycle of exclusion is neither irreversible, nor inevitable."

Maybe one day we will see the "domestication of biotechnology," as envisaged by the physicist and writer Freeman Dyson. In his vision, the tools of innovation will be taken out of the hands of the giant corporations and given to small farmers, or, as Dyson put it, "into the explosion of diversity of new living creatures, rather than the tightly controlled monoculture crops that the big corporations enjoy."

Jefferson is perhaps less science fiction and more social traction. His vision is of an explosion of small to medium enterprises, using widely available lab tools and informed by thoughtful science, providing a robust and sustainable solution.

In the end, the debate over biotechnology in agriculture may well be settled, as Calestous Juma argues, not by the arguments of advocates and critics but by the practical effects of the range of useful products available to humanity. In the meantime, it's easy to take a position pro- or anti-Bill Gates, as long as you don't have to address the problem of the world's food supply. But the Gates era of philanthropy promises the first major investments in agricultural development in the developing world, including biotechnology in its fullest sense. Rather than simply decry

Gates as a corporate Trojan horse, the task, surely, is to steer the funds he provides into a more democratic, more open, less monopolistic approach that can help free the peoples of the developing world from food dependence, offering them not only food security but also food sovereignty. The goal should be a world in which free people not only have enough to eat but also have the power to determine for themselves where their food comes from, who produces it, and under what conditions it is grown.

# Another Take | HAZARDS OF GENETICALLY ENGINEERED FOODS AND CROPS
## WHY WE NEED A GLOBAL MORATORIUM

### By Ronnie Cummins, Organic Consumers Association

Representing more than 850,000 members, subscribers, and volunteers, including several thousand businesses in the natural foods and organic marketplace, the Organic Consumers Association (OCA) is the only organization in the United States focused exclusively on promoting the views and interests of the nation's estimated fifty million organic and socially responsible consumers. The OCA deals with crucial issues of food safety, industrial agriculture, genetic engineering, children's health, corporate accountability, fair-trade, environmental sustainability, and other key topics. You can learn more about the OCA and its work by visiting the organization's website at http://www.organicconsumers.org.

In this selection, OCA's national director offers a forceful presentation of the case against genetically modified foods, summarizing a host of arguments—scientific, medical, environmental, economic, and ethical—supporting the need for a moratorium on the development and use of GM organisms.

The technology of genetic engineering (GE) is the practice of altering or disrupting the genetic blueprints of living organisms—plants, trees, fish, animals, humans, and microorganisms. This technology is wielded by transnational "life science" corporations such as Monsanto and Aventis, who patent these blueprints and sell the resulting gene-foods, seeds, or other products for profit. Life science corporations proclaim that their new products will make agriculture sustainable, eliminate world hunger, cure disease, and vastly improve public health. However, these gene engineers have made it clear, through their business practices and political lobbying, that they intend to use GE to monopolize the global market for seeds, foods, fiber, and medical products.

GE is a revolutionary new technology that is still in its early experimental stages of development. This technology has the power to break down the natural genetic barriers—not only between species—but between humans, animals, and plants. Randomly inserting together the genes of nonrelated species—utilizing viruses, antibiotic-resistant genes, and bacteria as vectors, markers, and promoters—permanently alters their genetic codes. The gene-altered organisms that are created pass these genetic changes on to their offspring through heredity. Gene engineers all over the world are now snipping, inserting, recombining, rearranging, editing, and programming genetic material.

Animal genes and even human genes are randomly inserted into the chromosomes of plants, fish, and animals, creating heretofore unimaginable transgenic life forms. For the first time in history, transnational biotechnology corporations are becoming the architects and "owners" of life.

With little or no regulatory restraints, labeling requirements, or scientific protocol, bioengineers have begun creating hundreds of new GE "frankenfoods" and crops. The research is done with little concern for the human and environmental hazards and the negative socioeconomic impacts on the world's several billion farmers and rural villagers.

An increasing number of scientists are warning that current gene-splicing techniques are crude, inexact, and unpredictable—and therefore inherently dangerous. Yet, pro-biotech governments and regulatory agencies, led by the United States, maintain that GE foods and crops are "substantially equivalent" to conventional foods, and therefore require neither mandatory labeling nor premarket safety testing.

This Brave New World of frankenfoods is frightening. There are currently more than four dozen GE foods and crops being grown or sold in the United States. These foods and crops are widely dispersed into the food chain and the environment. More than eighty million acres of GE crops are presently under cultivation in the United States, while up to 750,000 dairy cows are being injected regularly with Monsanto's recombinant bovine growth hormone (rBGH).

Most supermarket processed food items now "test positive" for the presence of GE ingredients. In addition, several dozen more GE crops are in the final stages of development and will soon be released into the environment and sold in the marketplace. The "hidden menu" of these unlabeled GE foods and food ingredients in the United States now includes soybeans, soy oil, corn, potatoes, squash, canola oil, cottonseed oil, papaya, tomatoes, and dairy products.

GE food and fiber products are inherently unpredictable and dangerous—for humans, for animals, for the environment, and for the future of sustainable and organic agriculture. As Dr. Michael Antoniou, a British molecular scientist, points out, gene-splicing has already resulted in the "unexpected production of toxic substances . . . in genetically engineered bacteria, yeast, plants, and animals with the problem remaining undetected until a major health hazard has arisen." The hazards of GE foods and crops fall into three categories: human health hazards, environmental hazards, and socioeconomic hazards. A brief look at the already proven and likely hazards of GE products provides a convincing argument for why we need a global moratorium on all GE foods and crops.

## TOXINS AND POISONS

GE products clearly have the potential to be toxic and a threat to human health. In 1989, a genetically engineered brand of l-tryptophan, a common dietary supplement, killed thirty-seven Americans. More than 5,000 others were permanently disabled or afflicted with a potentially fatal and painful blood disorder, eosinophilia myalgia syndrome (EMS), before it was recalled by the Food and Drug Administration (FDA). The manufacturer, Showa Denko, Japan's third largest chemical company, had for the first time in 1988–1989 used GE bacteria to produce the over-the-counter supplement. It is believed that the bacteria somehow became contaminated during the recombinant DNA process. Showa Denko has paid out over $2 billion in damages to EMS victims.

In 1999, front-page stories in the British press revealed Rowett Institute scientist Dr. Arpad Pusztai's explosive research findings that GE potatoes are poisonous to mammals. These potatoes were spliced with DNA from the snowdrop plant and a commonly used viral promoter, the cauliflower mosaic virus (CAMV). GE snowdrop potatoes were found to be significantly different in chemical composition from regular potatoes and when fed to lab rats, damaged their vital organs and immune systems. The damage to the rats' stomach linings apparently was caused by the CAMV viral promoter, apparently giving the rats a severe viral infection. Most alarming of all, the CAMV viral promoter is spliced into nearly all GE foods and crops.

Dr. Pusztai's path-breaking research work unfortunately remains incomplete. Government funding was cut off, and he was fired after he spoke to the media. More and more scientists around the world are warning that genetic manipulation can increase the levels of natural plant toxins or allergens in foods (or create entirely new toxins) in unexpected ways by switching on genes that produce poisons. Since regulatory agencies do not currently require the kind of thorough chemical and feeding tests that Dr. Pusztai was conducting, consumers have now become involuntary guinea pigs in a vast genetic experiment. Dr. Pusztai warns, "Think of William Tell shooting an arrow at a target. Now put a blindfold on the man doing the shooting and that's the reality of the genetic engineer doing a gene insertion."

## INCREASED CANCER RISKS

In 1994, the FDA approved the sale of Monsanto's controversial rBGH. This GE hormone is injected into dairy cows to force them to produce more milk. Scientists have warned that significantly higher levels (400–500 percent or more) of a potent chemical hormone, insulin-like growth factor (IGF-1), in the milk and dairy products of rBGH injected cows could pose serious hazards such as human breast, prostate, and colon cancer. A number of studies have shown that humans with elevated levels of IGF-1 in their bodies are much more likely to get cancer. The U.S. congressional watchdog agency, the Government Accountability Office (GAO), told the FDA not to approve rBGH. They argued that injecting the cows with rBGH caused higher rates of udder infections, requiring increased antibiotic treatment. The increased use of antibiotics poses an unacceptable risk for public health. In 1998, Monsanto/FDA documents that had previously been withheld were released by government scientists in Canada showing damage to laboratory rats fed dosages of rBGH. Significant infiltration of rBGH into the prostate of the rats as well as thyroid cysts indicated potential cancer hazards from the drug. Subsequently, the government of Canada banned rBGH in early 1999. The European Union (EU) has had a ban in place since 1994. Although rBGH continues to be injected into ten percent of all U.S. dairy cows, no other industrialized country has legalized its use. The GATT Codex Alimentarius, a United Nations food standards body, has refused to certify that rBGH is safe.

## FOOD ALLERGIES

In 1996, a major GE food disaster was narrowly averted when Nebraska researchers learned that a Brazil nut gene spliced into soybeans could induce potentially fatal allergies in people sensitive to Brazil nuts. Animal tests of these Brazil nut-spliced soybeans had turned up negative. People with food allergies (which currently afflict eight percent of all American children), whose symptoms can range from mild unpleasantness to sudden death, may likely be harmed by exposure to foreign proteins spliced into common food products. Since humans have never before eaten most of the foreign proteins now being gene-spliced into foods, stringent pre-market safety testing (including long-term animal feeding and volunteer human feeding studies) is necessary in order to prevent a future public health disaster. Mandatory labeling is also necessary so that those suffering from food allergies can avoid hazardous GE foods and so that public health officials can trace allergens back to their source when GE-induced food allergies break out.

In fall 2001, public interest groups, including Friends of the Earth and the Organic Consumers Association, revealed that lab tests indicated that an illegal and likely allergenic variety of GE, Bt-spliced corn called StarLink had been detected in Kraft Taco Bell shells, as well as many other brand name products. The StarLink controversy generated massive media coverage and resulted in the recall of hundreds of millions of dollars of food products and seeds.

## DAMAGE TO FOOD QUALITY AND NUTRITION

A 1999 study by Dr. Marc Lappé published in the *Journal of Medicinal Food* found that concentrations of beneficial phytoestrogen compounds thought to protect against heart disease and cancer were lower in GE soybeans than in traditional strains. These and other studies, including Dr. Pusztai's, indicate that GE food will likely result in foods lower in quality and nutrition. For example, the milk from cows injected with rBGH contains higher levels of pus, bacteria, and fat.

## ANTIBIOTIC RESISTANCE

When gene engineers splice a foreign gene into a plant or microbe, they often link it to another gene, called an antibiotic resistance marker gene (ARM), that helps determine if the first gene was successfully spliced into the host organism. Some researchers warn that these ARM genes might unexpectedly recombine with disease-causing bacteria or microbes in the environment or in the guts of animals or people who eat GE food. These new combinations may be contributing to the growing public health danger of antibiotic resistance—of infections that cannot be cured with traditional antibiotics, for example, new strains of salmonella, E. coli, camphylobacter, and enterococci. German researchers have found antibiotic-resistant bacteria in the guts of bees feeding on gene-altered rapeseed (canola) plants. EU authorities are currently considering a ban on all GE foods containing antibiotic-resistant marker genes.

## INCREASED PESTICIDE RESIDUES

Contrary to biotech industry propaganda, recent studies have found that U.S. farmers growing GE crops are using just as many toxic pesticides and herbicides as conventional farmers and in some cases are using more. Crops genetically engineered to be herbicide-resistant account for almost eighty percent of all GE crops planted in 2000. The "benefits" of these herbicide-resistant crops are that farmers can spray as much of a particular herbicide on their crops as they want—killing the weeds without damaging their crop. Scientists estimate that herbicide-resistant crops planted around the globe will triple the amount of toxic broad-spectrum herbicides used in agriculture. These broad-spectrum herbicides are designed to literally kill everything green. The leaders in biotechnology are the same giant chemical companies—Monsanto, DuPont, Aventis, and Syngenta (the merger between Novartis and Astra-Zeneca)—that sell toxic pesticides. The same companies that create the herbicide-resistant GE plants are also selling the herbicides. The farmers are then paying for more herbicide treatment from the same companies that sold them the herbicide-resistant GE seeds.

## GENETIC POLLUTION

"Genetic pollution" and collateral damage from GE field crops already have begun to wreak environmental havoc. Wind, rain, birds, bees, and insect pollinators have begun carrying genetically altered pollen into adjoining fields, polluting the DNA of crops of organic and non-GE farmers. An organic farm in Texas has been contaminated with genetic drift from GE crops grown on a nearby farm. EU regulators are considering setting an "allowable limit" for genetic contamination of non-GE foods because they don't believe genetic pollution can be controlled.

Because they are alive, gene-altered crops are inherently more unpredictable than chemical pollutants—they can reproduce, migrate, and mutate. Once released, it is virtually impossible to recall GE organisms back to the laboratory or the field.

## DAMAGE TO BENEFICIAL
## INSECTS AND SOIL FERTILITY

In 1999, Cornell University researchers made a startling discovery. They found that pollen from GE Bt corn was poisonous to Monarch butterflies. The study adds to a growing body of evidence that GE crops are adversely affecting a number of beneficial insects, including ladybugs and lacewings, as well as beneficial soil microorganisms, bees, and possibly birds.

## CREATION OF GE
## "SUPERWEEDS" AND "SUPERPESTS"

Genetically engineering crops to be herbicide-resistant or to produce their own pesticide presents dangerous problems. Pests and weeds will inevitably emerge that are pesticide- or herbicide-resistant, which means that stronger, more toxic chemicals will be needed to get rid of the pests.

Herbicide-resistant "superweeds" are already emerging. GE crops such as rapeseed (canola) have spread their herbicide-resistance traits to related weeds, such as wild mustard plants. Lab and field tests also indicate that common plant pests such as cotton bollworms, living under constant pressure from GE crops, will soon evolve into "superpests" completely immune to Bt sprays and other environmentally sustainable biopesticides. This will present a serious danger for organic and sustainable farmers whose biological pest management practices will be unable to cope with increasing numbers of superpests and superweeds.

## NEW VIRUSES AND PATHOGENS

Gene-splicing will inevitably result in unanticipated outcomes and dangerous surprises that damage plants and the environment. Several years ago, researchers conducting experiments at Michigan State University found that genetically altering plants to resist viruses can cause the viruses to mutate into new, more virulent forms. Scientists in Oregon found that a GE soil microorganism, Klebsiella planticola, completely killed essential soil nutrients. Environmental Protection Agency (EPA) whistle-blowers issued similar warnings in 1997, protesting government approval of a GE soil bacterium called Rhizobium melitoli.

## GENETIC "BIO-INVASION"

By virtue of their "superior" genes, some GE plants and animals will inevitably run amok, overpowering wild species in the same way that exotic species, such as kudzu vine and Dutch elm disease, have created problems when introduced in North America. What will happen to wild fish and marine species, for example, when scientists release into the environment carp, salmon, and trout that are twice as large, and eat twice as much food, as their wild counterparts?

## SOCIOECONOMIC HAZARDS

The patenting of GE foods and widespread biotech food production threatens to eliminate farming as it has been practiced for 12,000 years. GE patents, such as the Terminator Technology, will render seeds infertile and force hundreds of millions of farmers who now save and share their seeds to purchase ever-more expensive GE seeds and chemical inputs from a handful of global biotech/seed monopolies. If the trend is not stopped, the patenting of transgenic plants and food-producing animals will soon lead to universal "bioserfdom" in which farmers will lease their plants and animals from biotech conglomerates such as Monsanto and pay royalties on seeds and offspring. Family and indigenous farmers will be driven off the land, and consumers' food choices will be dictated by a cartel of transnational corporations. Rural communities will be devastated. Hundreds of millions of farmers and agricultural workers worldwide will lose their livelihoods.

## ETHICAL HAZARDS

The genetic engineering and patenting of animals reduce living beings to the status of manufactured products. A purely reductionist science, biotechnology reduces all life to bits of information (genetic code) that can be arranged and rearranged at whim. Stripped of their integrity and sacred qualities, animals that are merely objects to their "inventors" will be treated as such. Currently, hundreds of GE "freak" animals are awaiting patent approval from the federal government. One can only wonder, after the wholesale gene altering and patenting of animals, will GE "designer babies" be next?

## WHAT CAN YOU DO?
## GUIDELINES FOR LOCAL GRASSROOTS ACTION

*Campaign Goals.*    As the anti-GE campaign in Europe has shown, mass grassroots action is the key to stopping this technology and moving agriculture in an organic and sustainable direction. The OCA advocates the following Food Agenda 2000–2010 as the foundation for our local-to-global campaign work:

- *Establish a Global Moratorium on All Genetically Engineered Foods and Crops.* These products have not been proven safe for human health and the environment, and they must be taken off the market.

- *Stop Factory Farming.* Begin the phaseout of industrial agriculture and factory farming—with a goal of significantly reducing the use of toxic chemicals and animal drugs on conventional farms by the year 2010. This phaseout will include a ban on the most dangerous farm chemicals and animal feed additives (antibiotics, hormones, and rendered animal protein) as well as the implementation of intensive integrated pest management practices (decrease the use of toxic pesticides and chemical fertilizers through natural composting, crop rotation, cover crops, use of beneficial insects, etc.).
- *Convert American Agriculture to at Least Thirty Percent Organic by the Year 2010.* We demand government funding and implementation of transition to organic programs so that at least thirty percent of U.S. agriculture is organic by the year 2010—with a strong emphasis on production for local and regional markets by small- and medium-sized organic farmers.

## TAKE ACTION IN YOUR COMMUNITY

- Circulate our Food Agenda 2000–2010 petition to identify as many people as possible who support our campaign. We will include these names in our local databases for two-way communication and mobilization.
- Help us find retail stores and co-ops that will circulate our petitions and other materials, which can all be downloaded from our website.
- Tell your friends and family about our free electronic newsletter, *BioDemocracy News.*
- Tune in to our OCA website for regular news, updates, and action alerts: www.organicconsumers.org.
- Organize forums, protests, and news-making events.
- Pressure elected public officials, political candidates, and regulatory agencies both locally and nationally to demand either an outright GE moratorium or comprehensive mandatory labeling, stringent pre-market safety testing, and long-term liability insurance for all GE food and fiber products.
- Contact us and support this campaign by sending a tax-deductible donation and/or volunteer to help with grass roots organizing. Contact us at:

Organic Consumers Association
6101 Cliff Estate Road
Little Marais, MN 55614
Phone: 218-226-4164; fax: 218-226-4157
Email: campaign@organicconsumers.org

Or make your donation via the website at: www.organicconsumers.org.

FIVE

# THE ETHANOL SCAM

## *BURNING FOOD TO MAKE MOTOR FUEL*

### *By Robert Bryce*

Robert Bryce is an Austin, Texas-based journalist who has been writing about the energy sector for nearly two decades. His articles have appeared in dozens of publications, including the *Atlantic Monthly*, *Slate*, *The New York Times*, *The Washington Post*, *The American Conservative*, *The Nation*, *The Washington Spectator*, and *The Guardian*. His first book, *Pipe Dreams: Greed, Ego, and the Death of Enron*, received rave reviews and was named one of the best nonfiction books of 2002 by *Publishers Weekly*. His second book, *Cronies: Oil, the Bushes, and the Rise of Texas, America's Superstate*, was published in 2004. His third book, *Gusher of Lies: The Dangerous Delusions of "Energy Independence,"* published in March 2008, was favorably reviewed by more than twenty media outlets and was praised by *American* magazine as "a strong and much-needed dose of reality."

Bryce spent twelve years writing for the *Austin Chronicle*. He now works as the managing editor of Houston-based magazine *Energy Tribune*. He is also a contributing writer at the *Texas Observer*.

The huge corn ethanol mandates imposed by Congress in the early years of this century may be the single most misguided agricultural program in modern American history.

That's saying something. But consider a few of the many harmful impacts of the ethanol mandates: higher food prices, increased air pollution from burning ethanol-flavored gasoline, increased water consumption, and increased water pollution. Plus, there's strong evidence that more

greenhouse gases are produced during the corn ethanol production process than are made during the production of conventional gasoline.

Why then, given these many problems, hasn't Congress rolled back
the ethanol mandates? The answer can be boiled down to a few salient
realities of American politics and agricultural policy.

First, the power wielded in Washington by lobbyists from the farm
states remains enormous. And within the subsidy-rich world of U.S.
agriculture, corn is king. What's more, Iowa is ground zero for corn. The
state holds the first presidential primary. The Iowa Caucuses set the tone
for the quadrennial national elections. And that forces even supposed
change agents such as the 2008 presidential candidates, Barack Obama
and John McCain, to genuflect before the altar of corn ethanol.

And, finally, once a juggernaut like the corn ethanol industry gets
rolling with massive federal support and mandated production levels,
bringing it to a halt is enormously difficult—even when study after
study shows that relying on corn ethanol as a cornerstone of an alleged
renewable energy policy is folly.

The corn ethanol program is a case study in how political pressure,
the power of giant agribusinesses, and willful ignorance of basic science
can be used to distort U.S. farm policy, producing results that include
not only economic damage to the nation but the potential for increasing
hunger around the world, especially among the most vulnerable people
on our planet.

———————

Of the many outrages of the corn ethanol program, surely the most
shameful aspect is this: Congress has mandated the creation of—and the
payment of subsidies to—a multibillion-dollar network of distilleries
that are burning food to make motor fuel at a time when there is a growing global shortage of food and *no* shortage of motor fuel.

As Lester Brown aptly put it in a piece published by *The Washington
Post* in 2006, these federal mandates are "setting the stage for an epic
competition . . . between the world's supermarkets and its service stations."[1] Indeed, the United States has created a corn-gobbling monster
that links the price of food with the price of oil. And it has done so even

though corn ethanol will never supply more than single-digit percentages of America's overall oil needs.

Even if all the corn grown in America were turned into ethanol, it would supply less than six percent of America's total oil needs.[2] The same limits face biodiesel made from soybeans. Even if all of the soybeans grown in America were converted into biodiesel, it would only supply about 1.5 percent of America's total annual oil needs.[3]

Of course, sensible policy would favor feeding people, not automobiles. But by hyping the dangers of foreign oil and by promoting the dangerous delusion of "energy independence," the farm lobby, working closely with legislative delegations from the farm states, convinced Congress that biofuels should be a key element of U.S. energy policy. The result: in 2005 and again in 2007, Congress passed laws mandating increasing use of corn ethanol in America's motor fuel mix. Those mandates have led to a flood of capital investment in the corn ethanol sector. America now has about 200 corn distilleries, built at a cost of more than $15 billion,[4] that will be capable of producing some 13.75 billion gallons of corn ethanol per year.[5] And those plants will be gobbling up about one-third of all the corn grown in the United States, about 4.1 billion bushels per year. That's twice the volume of corn consumed by the ethanol industry in 2006, seven times the quantity used in 2000, and nearly thirteen times the amount used in 1990.[6]

Need another comparison? That 4.1 billion bushels of corn being used to make ethanol is more than twice as much corn as was produced by the entire European Union in 2006 and more than five times as much as was raised in Mexico.[7]

This increasing use of food to make motor fuel poses a moral question: what should the United States be doing to help feed the growing ranks of the world's poor? As Brown noted in his *Washington Post* essay, "The grain required to fill a 25-gallon SUV gas tank with ethanol would feed one person for a full year."[8] And yet the United States is providing huge subsidies to a program that feeds cars, not people. At the same time, the corn ethanol business is causing a myriad of other problems, including worsening air quality, depletion of our water resources, pollution of our water resources, and worsening greenhouse gas emissions, and that's just a partial list.

The ethanol mandates are the essence of fiscal insanity: Congress is requiring the production of subsidized motor fuel from corn, the most-subsidized crop in America. According to the Environmental Working Group, between 1995 and 2006 federal corn subsidies, which are provided through myriad programs, totaled $56.1 billion. That's more than twice the amount given to any other commodity, including American mainstays like wheat and cotton, and 105 times more than was paid to tobacco farmers.[9]

While the environmental and fiscal effects are important, the key issue is food. The ethanol scam is contributing to the rise in global grain prices, which will raise food costs for some of the world's most vulnerable people.

The United States is setting the table for what could be a food disaster. The expansion of the ethanol business is occurring at the same time that global grain reserves are shrinking, world population is increasing, and global agricultural productivity is slowing. In July 2008, the U.S. Department of Agriculture (USDA) released a report that showed that global grain reserves were at their lowest levels since 1970. Furthermore, those grain reserves—about 300 million tons—were less than half the reserve volumes on hand as recently as 1997.[10] And the agency expects the downward trend to continue. In fact, it expects that U.S. stocks of corn, wheat, and soybeans will be at or near historic lows through 2016.[11] In a 2007 report, it also plainly stated that ethanol production is to blame for the lower grain reserves, saying that "Strong ethanol demand sharply lowers U.S. corn stocks in the projections."[12]

What will low stocks of corn mean at a time of increasing ethanol production? According to USDA economist Paul Westcott, it will mean greater price volatility. In September 2007, Westcott wrote a piece for *Amber Waves*, an agency magazine, in which he said that as ethanol production increases, demand for corn will increase and that demand will be "unresponsive to price changes." That is, the owners of the distilleries—who have invested tens of millions of dollars in their plants—will be willing to pay drastically higher prices for corn in order to continue operating.[13] This inelastic demand for corn, coupled with higher prices and lower stocks, "will make the corn market more vulnerable to shocks, such as production shortfalls due to weather, pests, or other factors," wrote

**Figure 5.1    U.S. Stocks-to-Use Ratios, 1980 to 2016: Corn, Wheat, and Soybeans**

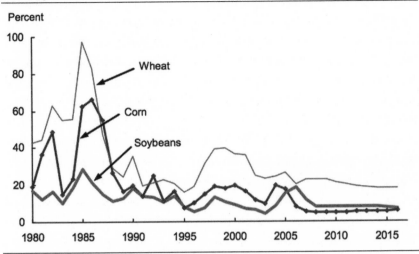

SOURCE: U.S. Department of Agriculture, "USDA Long-Term Projections, February 2007," 31. Available online at http://www.ers.usda.gov/publications/oce071/oce20071c.pdf.

Westcott. "As demand for corn becomes more inelastic, a greater change in market prices would be needed in response to a shock to bring the market to equilibrium. Thus, overall price variability and market volatility in the agricultural sector are likely to increase."[14]

Price variability and market volatility for food commodities have the greatest effect on low-income consumers, most of whom don't have the financial resources to absorb sharp increases in prices. That price volatility will likely have the biggest effect on the world's newest residents, most of whom will be born in developing countries. The world population, now at about 6.7 billion, is expected to increase by about one percent per year, to some 8.2 billion by 2030.[15]

Adding yet more urgency to the problem is the slowdown in farm yields. In July 2008, the USDA reported that it expects agricultural productivity between 2009 and 2017 to increase at about half the rate that prevailed between 1970 and 1990. The agency reported that between 1970 and 1990, "production rose an average 2.2 percent per year. Since 1990, the growth rate has declined to about 1.3 percent. USDA's 10-year agricultural projections for U.S. and world agriculture see the rate declining to 1.2 percent per year between 2009 and 2017."[16]

These trends are occurring at the same time that Congress has created a direct link between the price of gasoline and the price of corn that will be difficult, if not impossible, to break in the near future. Here's why: even if Congress repeals the mandates on ethanol production (by 2015, the United States must be using 15 billion gallons of ethanol in its gasoline) and eliminates the subsidies for ethanol production, the industry will not shut down. Even without federal supports, many of the 200 distilleries will still be profitable. And their profitability will be directly linked to the price of oil. As the price of oil rises, some of the more efficient ethanol producers will be able to produce unsubsidized corn ethanol at a profit because their fuel will be competing with higher-priced gasoline for market share. That means that big oil refiners will likely decide to blend increasing quantities of less-expensive corn ethanol with their gasoline in order to reduce their costs.

Before going further into the ethanol mess, let me stipulate the obvious. Several factors have helped drive grain prices higher over the past few years, including the falling value of the dollar, increasing energy prices, increasing global demand for grains and meat, and poor grain harvests in some countries. That said, it's also clear that the ethanol mandates are a key, and easily avoidable, factor in the food inflation equation.

Over the past two years or so, at least eleven reports have determined that ethanol production is driving up food prices. For instance, in March 2008, a report commissioned by the Coalition for Balanced Food and Fuel Policy (a coalition based in Washington, D.C., of eight meat, dairy, and egg producers' associations) estimated that the biofuels mandates passed by Congress will cost the U.S. economy more than $100 billion from 2006 to 2009. The report declared that "[t]he policy favoring ethanol and other biofuels over food uses of grains and other crops acts as a regressive tax on the poor." It went on to estimate that the total cost of the U.S. biofuels mandates will total some $32.8 billion this year, or about $108 for every American citizen.[17]

An April 8, 2008, internal report by the World Bank found that grain prices increased by 140 percent between January 2002 and February 2008. According to the report, "[t]his increase was caused by a confluence of factors but the most important was the large increase in biofuels production in the U.S. and E.U. Without the increase in biofuels, global wheat and maize [corn] stocks would not have declined appreciably and price increases due to other factors would have been moderate."[18] Robert Zoellick, president of the bank, acknowledged those facts, saying that biofuels are "no doubt a significant contributor" to high food costs. And he said that "it is clearly the case that programs in Europe and the United States that have increased biofuel production have contributed to the added demand for food."[19]

In May 2008, the Congressional Research Service blamed recent increases in global food prices on two factors: increased grain demand for meat production and the biofuels mandates. The agency said that the recent "rapid, 'permanent' increase in corn demand has directly sparked substantially higher corn prices to bid available supplies away from other uses—primarily livestock feed. Higher corn prices, in turn, have forced soybean, wheat, and other grain prices higher in a bidding war for available crop land."[20]

That same month, Mark W. Rosegrant of the International Food Policy Research Institute, a Washington, D.C.-based think tank whose vision is "a world free of hunger and malnutrition," testified before the U.S. Senate on biofuels and grain prices.[21] Rosegrant said that the ethanol mandates caused the price of corn to increase by twenty-nine percent, rice to increase by twenty-one percent, and wheat by twenty-two percent.[22] Rosegrant estimated that if the global biofuels mandates were eliminated altogether, corn prices would drop by twenty percent, while sugar and wheat prices would drop by eleven percent and eight percent, respectively, by 2010.[23]

Rosegrant added that "[i]f the current biofuel expansion continues, calorie availability in developing countries is expected to grow more slowly; and the number of malnourished children is projected to increase." He continued, saying, "It is therefore important to find ways to keep biofuels from worsening the food-price crisis. In the short run, removal of ethanol blending mandates and subsidies and ethanol import tariffs, and in the United States—together with removal of policies in Europe promoting biofuels—would contribute to lower food prices."[24]

In early July 2008, Britain's Renewable Fuels Agency concluded, "Biofuels contribute to rising food prices that adversely affect the poorest."[25] The report, known as the Gallagher Review, also said that demand for "[biofuels] production must avoid agricultural land that would otherwise be used for food production. This is because the displacement of existing agricultural production, due to biofuel demand, is accelerating land-use change and, if left unchecked, will reduce biodiversity and may even cause greenhouse gas emissions rather than savings. The introduction of biofuels should be significantly slowed."

Even the USDA, the federal agency that has long been one of the corn ethanol sector's biggest boosters, is admitting that corn ethanol is driving up food prices. That's somewhat remarkable given that the agency's leaders have consistently downplayed the link.[26] Nevertheless, in July 2008 the department released a report called "Food Security Assessment, 2007," which states very clearly that the biofuels mandates are pushing up food prices. The first page of the report says:

> the persistence of higher oil prices deepens global energy security concerns and heightens the incentives to expand production of other sources of energy including biofuels. The use of food crops for producing biofuels, growing demand for food in emerging Asian and Latin American countries, and unfavorable weather in some of the largest food-exporting countries in 2006–07 all contributed to growth in food prices in recent years.[27]

While that admission is noteworthy, the July 2008 report's importance lies with its projections about the growing numbers of people around the world who are facing food insecurity. And while the USDA report does not correlate this increasing food insecurity with soaring ethanol production, the connections are abundantly clear: as the United States uses more corn to make motor fuel, there is less grain available on the market. That means higher prices. And that's a key factor for residents of poor countries who generally spend a higher percentage of their income on food than their counterparts in the developed world.

For instance, in the United States only about 6.5 percent of disposable income is spent on food. By contrast, in India, about 40 percent of personal disposable income is spent on food. In the Philippines, it's about 47.5 percent.[28] In some sub-Saharan Africa countries, consumers spend about 50 percent of the household budget on food. And according to the USDA, "In some of the poorest countries in the region such as Madagascar, Tanzania, Sierra Leone, and Zambia, this ratio is more than 60 percent."[29]

The July 2008 USDA report goes on to say that the number of people facing food insecurity jumped from 849 million in 2006 to 982 million in 2007. And those numbers are expected to continue rising. By 2017, the number of food-insecure people is expected to hit 1.2 billion. And, says the USDA, "short-term shocks, natural as well as economic," could make the problem even worse.

Despite these many reports, the ethanol apologists continue to claim that their favorite fuel is not to blame. For instance, the Renewable Fuels Association, a U.S. trade group funded by the ethanol producers, has claimed that "corn demand for ethanol has no noticeable impact on retail food prices."[30] In May 2008, Charles Grassley, a Republican Senator from Iowa, gave a speech on the Senate floor, in which he denounced the "scapegoating" of ethanol as a cause of rising food prices. He declared that "[n]one of the criticisms are based on sound science, sound economics, or for that matter, even common sense."[31]

But the deleterious effects of the ethanol mandates go far beyond the supermarket aisle.

Clean air advocates contend that the increasing use of ethanol in gasoline is increasing the amount of smog in America's cities. William Becker, executive director of the National Association of Clean Air Agencies, which represents air pollution control authorities from forty-nine states and several territories as well as local agencies from 165 metro areas around the United States, told me that Congress "decided to mandate ethanol without first analyzing the air quality impacts." Gasoline that has been blended with ten percent ethanol may be more volatile than conventional gasoline, which means more light hydrocarbons can be emitted into the air. In addition, ethanol-blended gasoline is more likely to seep through the seals and gaskets in the engine (a process known as permeation), which also puts more hydrocarbons into the atmosphere. The increase in airborne hydrocarbons often means increases in key pollutants such as ground-level ozone. For Becker, the conclusion is crystal clear: "More ethanol means more air pollution. Period."

A spate of studies have shown that the production of corn ethanol likely creates more greenhouse gases than conventional gasoline. In August

2008, a study coauthored by Paul Crutzen, a Nobel Prize–winning chemist, found that due to releases of nitrous oxide during the production process, "commonly used biofuels such as biodiesel from rapeseed and bioethanol from corn (maize) can contribute as much or more to global warming" as fossil fuels.[32] That study confirms the findings of a 1997 report by the Government Accountability Office, which determined that corn ethanol production produces "relatively more nitrous oxide and other potent greenhouse gases. In contrast, the greenhouse gases released during the conventional gasoline fuel cycle contain relatively more of the less potent type, namely, carbon dioxide."[33]

In January 2008, researchers at the University of California at Berkeley published a paper that found that producing corn ethanol from land that was formerly held in the Conservation Reserve Program had greenhouse gas emissions that were 2.4 times greater than those from conventional gasoline.[34] In February 2008, *Science* magazine published a study that found that when accounting for land-use changes, corn ethanol production "nearly doubles greenhouse emissions over 30 years and increases greenhouse gases for 167 years." They also found that "biofuels from switchgrass, if grown on U.S. corn lands, increase emissions by 50 percent."[35]

Corn ethanol production has negative impacts on water quality. In late July 2008, researchers from the National Oceanic and Atmospheric Administration (NOAA), Louisiana State University, and the Louisiana Universities Marine Consortium reported that the "dead zone" in the Gulf of Mexico, a patch of ocean containing water with oxygen levels low enough to kill marine life, covered nearly 8,000 square miles. It is the second largest dead zone recorded since measurements began in 1985.[36] Since 1990, the dead zone has covered an average of 4,800 square miles.[37] A key reason for the increasing size of the dead zone is the increased planting of corn to meet the soaring demand from ethanol distilleries. That additional acreage has resulted in increased applications of fertilizers such as nitrogen and phosphorus, which are then being washed downstream.

Indeed, the ethanol sector's impact on water supplies may be one of the most important—and yet overlooked—aspects of the corn ethanol business. In December 2006, scientists at Sandia National Laboratory in New Mexico issued an eighty-page report called "Energy Demands on

Water Resources."[38] In a section on biofuels, the report says that the amount of water needed to grow corn varies widely, but it states that the amount of "water required for production of irrigated corn is 11,000 gal per MMBtu."[39] In layperson's terms, that means that ethanol produced from irrigated corn requires 11,000 gallons per million Btus of fuel produced, or about 880 gallons of water for every gallon of ethanol. Add in another five gallons of water needed at the distillery to turn the corn into ethanol, and each gallon of ethanol requires 885 gallons of fresh water.[40] For comparison, the average bathtub holds thirty-five gallons.[41] Thus, a gallon of ethanol produced from irrigated corn requires as much water as the amount contained in twenty-five bathtubs.

Of course, not all corn is irrigated. By some estimates, just fifteen percent of all U.S. corn is produced with irrigation.[42] Even so, local demands for water to produce corn ethanol can be extraordinary. For instance, some seventy percent of the corn grown in Nebraska—the third largest corn producing state—relies on irrigation.[43]

But even if only fifteen percent of the corn in the United States is produced with irrigation, ethanol production is still hugely water-intensive. The math is clear: assume that fifteen percent of the corn used during the production of America's ethanol in 2006 came from irrigated fields. In that case, each gallon of domestic ethanol required the consumption of about 132 gallons of water.[44] That's a huge quantity, particularly when compared to the quantity needed for oil and gas production. According to the Sandia report, the extraction and refining of conventional oil requires—at most—2.8 gallons of water for each gallon of oil produced.[45] (The report puts the minimum water consumption at about 1.3 gallons per gallon of refined oil product).

In June 2007, the results of the Sandia report were corroborated by two Colorado scientists, Jan F. Kreider, an emeritus engineering professor at the University of Colorado, and Peter S. Curtiss, a Boulder-based engineering consultant. In fact, they found that the Sandia estimates of ethanol's water needs may be too low. Kreider and Curtiss claim that each gallon of corn ethanol requires 170 gallons of water. (For comparison, they estimated that producing one gallon of gasoline requires about five gallons of water.) If Kreider and Curtiss are right, that means that the five billion gallons of corn ethanol that were produced in America in 2006 used more water than all that was used during the production of all of the gasoline consumed in the United States that year.[46]

But none of those externalities seem to matter to Congress or the special interest groups who are pushing for expansion of the biofuels mandates. The reason for that ignorance is largely about economics. The 200 or so distilleries now operating in the United States are spread among twenty-six states.[47] And members of the House and Senate want to keep those plants operating. That will likely mean a continuation of the primary ethanol subsidy, known as the Volumetric Ethanol Excise Tax Credit, which provides a tax credit of $0.51 for each gallon of ethanol that is blended into gasoline.[48] The voting power of the farm state delegations likely also assures the continuation of huge corn subsidies. And those subsidies do not appear to respond to changes in the size of the harvest. For instance, in 2006, the U.S. corn harvest was 10.5 billion bushels,[49] and federal corn subsidies totaled $4.9 billion.[50] In 2005, farmers enjoyed a larger crop, some 11.1 billion bushels,[51] but federal subsidies were almost twice as high, totaling almost $9.4 billion.[52] And the latest farm bill, a $307 billion behemoth passed by Congress in 2008, assures that those corn subsidies will continue for years to come.

The punch line here is obvious: In U.S. agriculture, Big Corn is king. Big Corn now rules on nearly one-fourth of all the acreage under cultivation in America and, thanks to congressional mandates and subsidies, it will continue burning food to make motor fuel for years to come. And that policy will likely mean increasing hunger for millions of the world's poorest people.

In the years to come, Americans can only hope that sanity—and, yes, morality—will prevail over politics, and that citizens and government leaders who are more concerned about the welfare of our nation and the world will come together to reform the massive U.S. farm subsidy system. And the first item on that reform list should be the most egregiously destructive program of all—the corn ethanol scam.

# Another Take | EXPOSURE TO PESTICIDES
## | A FACT SHEET

### By the Organic Consumers Association

As noted on page 79, the Organic Consumers Association (OCA) is the only organization in the United States focused exclusively on promoting the views and interests of the nation's estimated fifty million organic and socially responsible consumers. In this fact sheet, OCA provides some basic data about the exposure to pesticides experienced by ordinary Americans and the potential health risks it can cause.

The majority of this data, assembled in July 2005, comes from U.S. government agencies and their respective reports.

According to the Environmental Protection Agency (EPA) and the National Academy of Sciences, standard chemicals are up to ten times more toxic to children than to adults, depending on body weight. This is due to the fact that children take in more toxic chemicals relative to body weight than adults and have developing organ systems that are more vulnerable and less able to detoxify such chemicals.[1] According to the EPA's "Guidelines for Carcinogen Risk Assessment," children receive fifty percent of their lifetime cancer risks in the first two years of life.[2]

According to the Food and Drug Administration, half of produce currently tested in grocery stores contains measurable residues of pesticides. Laboratory tests of eight industry-leader baby foods reveal the presence of sixteen pesticides, including three carcinogens.[3] In blood samples of children aged two to four, concentrations of pesticide residues are six times higher in children eating conventionally farmed fruits and vegetables compared with those eating organic food.[4]

According to the U.S. Department of Health and Human Services, organophosphate pesticides (OP) are now found in the blood of ninety-five percent of Americans tested. OP levels are twice as high in blood samples taken from children than in adults. Exposure to OPs is linked to hyperactivity, behavior disorders, learning disabilities, developmental delays, and motor dysfunction. OPs account for half of the insecticides used in the United States.[5]

The U.S. Centers for Disease Control reports that one of the main sources of pesticide exposure for U.S. children comes from the food they eat.[6]

The U.S. Department of Agriculture strictly prohibits mixing different types of pesticides for disposal due to the well-known process of the individual chemicals combining into new, highly toxic chemical compounds. There are no regulations regarding pesticide mixture on a consumer product level even though, in a similar manner, those same individual pesticide residues interact and mix together into new chemical compounds when conventional multiple ingredient products are made. Sixty-two percent of food products tested contain a measurable mixture of residues of at least three different pesticides.[7]

Currently, more than 400 chemicals can be regularly used in conventional farming as biocides to kill weeds and insects. For example, apples can be sprayed up to sixteen times with thirty-six different pesticides. None of these chemicals are present in organic foods.[8]

More than 300 synthetic food additives are allowed by the FDA in conventional foods. None of these is allowed in foods that are USDA certified organic.

## SIX

# THE CLIMATE CRISIS AT
# THE END OF OUR FORK

### By Anna Lappé

---

Anna Lappé is a national best-selling author and public speaker known for her work on sustainable agriculture, food politics, and social change. Named one of *Time*'s Eco-Who's Who, Anna has been featured in *Gourmet, Food & Wine, The New York Times, Delicious Living,* and *O Magazine,* among many other outlets.

With her mother Frances Moore Lappé, Anna leads the Cambridge-based Small Planet Institute, a collaborative network for research and popular education, and the Small Planet Fund, which has raised more than half a million dollars for democratic social movements worldwide, two of which have won the Nobel Peace Prize since the fund's founding in 2002.

Anna's first book, *Hope's Edge: The Next Diet for a Small Planet* (Tarcher/Penguin, 2002), cowritten with Frances Moore Lappé, chronicles courageous social movements around the world addressing the root causes of hunger and poverty. Winner of the Nautilus Award for Social Change, *Hope's Edge* has been published in several languages and is used in classrooms across the country. She is also the coauthor of *Grub: Ideas for an Urban Organic Kitchen* (Tarcher/Penguin, 2006) with eco-chef Bryant Terry and is at work on her third book, *Eat the Sky* (Bloomsbury, 2010), about food, farming, and climate change.

Anna holds an MA in economic and political development from Columbia University's School of International and Public Affairs and graduated with honors from Brown University. From 2004 to 2006 she was a Food and Society Policy Fellow, a national program of the W. K. Kellogg Foundation. She lives in Fort Greene, Brooklyn, where she visits her local farmers' market as often as she can.

---

We could hear audible gasps from the two dozen New York state farmers gathered at the Glynwood Center on a cold December day in 2007 when NASA scientist Cynthia Rosenzweig, one of the world's leading experts on climate change and agriculture, explained the slide glowing on the screen in front of us.

The Glynwood Center, an education nonprofit and farm set on 225 acres in the Hudson Valley, had brought Rosenzweig to speak to area farmers about the possible impact of climate change on the region. Pointing to an arrow swooping south from New York, Rosenzweig said: "If we don't drastically reduce greenhouse gas emissions by 2080, farming in New York could feel like farming in Georgia.

"It was all projections before. It's not projections now—it's observational science," said Rosenzweig. We are already seeing major impacts of climate change on agriculture: droughts leading to crop loss and salinization of soils, flooding causing waterlogged soils, longer growing seasons leading to new and more pest pressures, and erratic weather shifting harvesting seasons, explained Rosenzweig.

When people think about climate change and food, many first think of the aspect of the equation that Rosenzweig focused on that day—the impact of climate change on farming. But when it comes to how the food system impacts global warming, most draw a blank.

Challenged to name the human factors that promote climate change, we typically picture industrial smokestacks or oil-thirsty planes and automobiles, not Pop-Tarts or pork chops. Yet the global system for producing and distributing food accounts for roughly *one-third* of the human-caused global warming effect. According to the United Nation's seminal report, *Livestock's Long Shadow*, the livestock sector alone is responsible for eighteen percent of the world's total global warming effect—more than the emissions produced by every plane, train, and steamer ship on the planet.[1]

Asked what we can do as individuals to help solve the climate change crisis, most of us could recite these eco-mantras from memory: Change our light bulbs! Drive less! Choose energy-efficient appliances! Asked what we can do as a nation, most of us would probably mention promoting renewable energy and ending our addiction to fossil fuels. Few among us would mention changing the way we produce our food or the dietary choices we make.

Unfortunately, the dominant storyline about climate change—its biggest drivers and the key solutions—diverts us from understanding how other sectors, particularly the food sector, are critical parts of the *problem*, but even more importantly can be vital strategies for *solutions*.

If the role of our food system in global warming comes as news to you, it's understandable. Many of us have gotten the bulk of our information about global warming from Al Gore's wake-up call *An Inconvenient Truth*, the 2006 Oscar-winning documentary that became the fourth-highest grossing nonfiction film in American history.[2] In addition to the record-breaking doc, Gore's train-the-trainer program, which coaches educators on sharing his slideshow, has further spread his central message about the threat posed by human-made climate change. But Gore's program offers little information about the connection between climate change and the food on your plate.

Mainstream newspapers in the United States haven't done a much better job of covering the topic. Researchers at Johns Hopkins University analyzed climate change coverage in sixteen leading U.S. newspapers from September 2005 through January 2008. Of the 4,582 articles published on climate change during that period, only 2.4 percent addressed the role of the food production system, and most of those only peripherally. In fact, just half of one percent of all climate change articles had "a substantial focus" on food and agriculture.[3] Internationally, the focus hasn't been much different. Until recently, much of the attention from the international climate change community and national coordinating bodies was also mostly focused on polluting industries and the burning of fossil fuels, not on the food sector.

This is finally starting to change. In the second half of 2008, writers from *O: The Oprah Magazine* to the *Los Angeles Times* started to cover the topic, increasing the public's awareness of the food and climate change connection. In September 2008, Dr. Rajendra Pachauri, the Indian economist serving his second term as chair of the United Nations Intergovernmental Panel on Climate Change, made a bold statement about the connection between our diet and global warming. Choosing to eat less meat, or eliminating meat entirely, is one of the most important personal choices we can make to address climate change, said Pachauri.[4] "In terms of immediacy of action and the feasibility of bringing about reductions in a short period of time, it clearly is the most attractive opportunity," said Pachauri. "Give up meat for one day [a week] initially, and decrease it from there."[5]

Why does our food system play such a significant role in the global warming effect? There are many reasons, including the emissions created by industrial farming processes, such as fertilizer production, and the carbon emissions produced by trucks, ships, and planes as they transport foods across nations and around the world. Among the main sources of the food system's impact on climate are land use changes, especially the expansion of palm oil production, and effects caused by contemporary agricultural practices, including the emissions produced by livestock.

## THE LAND USE CONNECTION

Let's look at land use first. A full eighteen percent of the world's global warming effect is associated with "land use changes," mostly from the food system.[6] The biggest factors are the destruction of vital rainforests through burning and clearing and the elimination of wetlands and peat bogs to expand pasture for cattle, feed crops for livestock, and oil palm plantations, especially in a handful of countries, Brazil and Indonesia chief among them.[7]

What do Quaker Granola Bars and Girl Scout Cookies have to do with the climate crisis?[8] These processed foods—along with other popular products, including cosmetics, soaps, shampoo, even fabric softeners—share a common ingredient, one with enormous climate implications: palm oil.[9] As the taste for processed foods skyrockets, so does the demand for palm oil, production of which has more than doubled in the last decade.[10] Today, palm oil is the most widely traded vegetable oil in the world, with major growth in the world's top two importing countries, India and China.[11]

As oil palm plantations expand on rainforests and peat lands in Southeast Asia, the natural swamp forests that formerly filled those lands are cut down and drained, and the peat-filled soils release carbon dioxide and methane into the atmosphere. (Methane is a key greenhouse gas with twenty-three times the global warming impact of carbon dioxide.) In a recent study, researchers estimate that producing one ton of palm oil can create fifteen to seventy tons of carbon dioxide over a twenty-five year period.[12]

Three of the world's biggest agribusiness companies are major players in the palm oil market, which is concentrated in two countries—Malaysia and Indonesia—where in 2007, forty-three percent and forty-four percent of the world's total palm oil was produced, respectively.[13] Wilmar, an affiliate of the multinational giant Archer Daniels Midland, is the largest palm oil producer in the world;[14] soy behemoth Bunge is a major importer of palm oil into the United States (although at the moment it doesn't own or operate any of its own facilities);[15] and grain-trading Cargill owns palm plantations throughout Indonesia and Malaysia.[16] These three companies and others producing palm oil claim that guidelines from the Roundtable on Sustainable Palm Oil (RSPO), established in 2004 by industry and international nonprofits, ensure sustainable production that minimizes the destruction of forest and peat bogs as well as deleterious effects on the global climate.[17]

However, some environmental and human rights groups argue that loopholes in the Roundtable's regulations still leave too much wiggle room. Says Greenpeace, "The existing standards developed by the RSPO will not prevent forest and peat land destruction, and a number of RSPO members are taking no steps to avoid the worst practices of the palm oil industry."[18]

We also know from new data that palm plantation expansion on peat land is not slowing. According to Dr. Susan Page from the University of Leicester, deforestation rates on peat lands have been increasing for twenty years, with one-quarter of all deforestation in Southeast Asia occurring on peat lands in 2005 alone.[19]

The other side of the land use story is deforestation driven by the increased production of livestock, expanding pasture lands and cropland for feed. In Latin America, for instance, nearly three-quarters of formerly forested land is now occupied by pastures; feed crops for livestock cover much of the remainder.[20] Globally, one-third of the world's arable land is dedicated to feed crop production.[21] Poorly managed pastures lead to overgrazing, compaction, and erosion, which release stored carbon into the atmosphere. With livestock now occupying twenty-six percent of the planet's ice-free land, the impact of this poor land management is significant.[22]

Raising livestock in confinement and feeding them diets of grains and other feedstock—including animal waste by-products—is a relatively recent phenomenon. In the postwar period, intensification of animal production was seen as the path to productivity. As livestock were confined in high stocking densities often far from where their feed was grown, a highly inefficient and environmentally costly system was born.

As a British Government Panel on Sustainable Development said in 1997, "Farming methods in the last half century have changed rapidly as a result of policies which have favored food production at the expense of the conservation of biodiversity and the protection of the landscape."[23] Despite these environmental costs, confined animal feeding operations (CAFOs) spread in the 1960s and 1970s into Europe and Japan and what was then the Soviet Union. Today, CAFOs are becoming increasingly common in East Asia, Latin America, and West Asia.

As the largest U.S.-based multinational meat companies, including Tyson, Cargill, and Smithfield, set their sights overseas, the production of industrial meat globally is growing.[24] In addition, the increasing supply of meat in developing countries flooded with advertising for Western-style eating habits is leading to a potential doubling in demand for industrial livestock production, and therefore feed crops, from 1997–1999 to 2030.[25]

Although the shift from traditional ways of raising livestock to industrial-scale confinement operations is often defended in the name of "efficiency," it's a spurious claim. As a way of producing edible proteins, feedlot livestock production is inherently inefficient. While ruminants such as cattle naturally convert inedible-to-humans grasses into high-grade proteins, under industrial production, grain-fed cattle pass along to humans only a fraction of the protein they consume.[26] Debates about this conversion rate abound. The U.S. Department of Agriculture estimates that it takes seven pounds of grain to produce one pound of beef.[27] However, journalist Paul Roberts, author of *The End of Food*, argues that the true conversion rate is much higher. While feedlot cattle need at least ten pounds of feed to gain one pound of live weight, Roberts states, nearly two-thirds of this weight gain is for inedible parts, such as bones, other organs, and hide. The true conversion ratio, Roberts estimates, is twenty pounds of grain to produce a single pound of beef, 7.3 pounds for pigs, and 3.5 pounds for poultry.[28]

The inefficiency of turning to grain-fed livestock as a major component of the human diet is devastating in itself, especially in a world where nearly one billion people still go hungry. But now we know there is a climate cost as well. The more consolidation in the livestock industry—where small-scale farmers are pushed out and replaced by large-scale confinement operations—the more land will be turned over to feed production. This production is dependent on fossil fuel–intensive farming, from synthesizing the human-made nitrogen fertilizer to using fossil fuel–based chemicals on feed crops. Each of these production steps cost in emissions contributing to the escalating greenhouse effect undermining our planet's ecological balance.

## THE AGRICULTURE CONNECTION

One reason we may have been slow to recognize the impact of the food system on climate change may be a certain "carbon bias." While carbon dioxide is the most abundant human-made greenhouse gases in the atmosphere, making up seventy-seven percent of the total human-caused global warming effect, methane and nitrous oxide contribute nearly all the rest.[29] (Other greenhouse gases are also relevant to the global warming effect, but are currently present in much smaller quantities and have a less significant impact.)[30] Agriculture is responsible for most of the human-made methane and nitrous oxide in the atmosphere, which contribute 13.5 percent of total greenhouse gas emissions, primarily from animal waste mismanagement, fertilizer overuse, the natural effects of ruminant digestion, and to a small degree rice production.[31] (1.5 percent of total emissions come from methane produced during rice cultivation).[32]

Though livestock only contribute nine percent of carbon dioxide emissions, the sector is responsible for thirty-seven percent of methane and sixty-five percent of nitrous oxide.[33] Here again, recent changes in agricultural practices are a significant factor. For centuries, livestock have been a vital part of sustainable food systems, providing muscle for farm work and meat as a vital protein source. Historically, properly grazed livestock produced numerous benefits to the land: hooves aerate soil, allowing more oxygen in the ground, which helps plant growth; their hoof action also presses grass seed into the earth, fostering plant growth, too; and, of course, their manure provides natural fertilizer. Indeed, new self-described "carbon farmers" are developing best management practices to manage cattle grazing to reduce compaction and overgrazing and, mimicking traditional grazing patterns, increasing carbon sequestration in the soil.[34]

But modern livestock production has steered away from these traditional practices toward the industrial-style production described above and to highly destructive overgrazing. In sustainable systems tapping nature's wisdom, there is no such thing as waste: manure is part of a holistic cycle and serves to fertilize the same lands where the animals that produce it live. In CAFOs, there is simply too much waste to cycle back through the system. Instead, waste is stored in manure "lagoons," as they're euphemistically called. Without sufficient oxygenation, this waste emits methane and nitrous oxide gas. As a consequence of industrial livestock production, the United States scores at the top of the world for methane emissions from manure. Swine production is king in terms of methane emissions, responsible for half of the globe's total.[35]

The sheer numbers of livestock exacerbate the problem. In 1965, eight billion livestock animals were alive on the planet at any given moment; ten billion were slaughtered every year. Today, thanks in part to CAFOs that spur faster growth and shorter lifespan, twenty billion livestock animals are alive at any moment, while nearly fifty-five billion are slaughtered annually.[36]

Ruminants, such as cattle, buffalo, sheep, and goats, are among the main agricultural sources of methane. They can't help it; it's in their nature. Ruminants digest through microbial, or enteric, fermentation, which produces methane that is then released by the animals, mainly through belching. While this process enables ruminants to digest fibrous grasses that we humans can't convert into digestible form, it also contributes to livestock's climate change impact. (Enteric fermentation accounts for twenty-five percent of the total emissions from the livestock sector; land use changes account for another 35.4 percent; manure accounts for 30.5 percent.)[37]

In addition to the ruminants' digestive process, emissions from livestock can be traced back to the production of the crops they consume. Globally, thirty-three percent of the world's cereal harvest and ninety percent of the world's soy harvest are now being raised for animal feed.[38] Feed crop farmers are heavily dependent on fossil fuels, used to power the on-farm machinery as well as used in the production of the petroleum-based chemicals to protect against pests, stave off weeds, and foster soil fertility on large-scale monoculture fields. In addition, these crops use up immense quantities of fertilizer. In the United States and Canada, half of all synthetic fertilizer is used for feed crops.[39] In the United Kingdom, the total is nearly seventy percent.[40] To produce this fertilizer requires tons of natural gas; on average 1.5 tons of oil equivalents are used up to make one ton of fertilizer.[41] Yet in the United States, only about half of the nitrogen fertilizer applied to corn is even used by the crop.[42] This needless waste is all the more alarming because nitrogen fertilizer contributes roughly three-quarters of the country's nitrous oxide emissions.

Erosion and deterioration of soils on industrial farms is another factor in the food sector's global warming toll. As industrial farms diminish natural soil fertility and disturb soil through tillage, soil carbon is released into the atmosphere.[43] Because industrialized agriculture also relies on huge amounts of water for irrigation, these farms will be more vulnerable as climate change increases drought frequency and intensity and decreases water availability. Globally, seventy percent of the world's available freshwater is being diverted to irrigation-intensive agriculture.[44]

## THE WASTE AND TRANSPORTATION CONNECTION

The sources of food system emissions on which we've focused so far—including land use changes and agricultural production—are responsible for nearly one-third of the total human-made global warming effect. That's already quite a lot, but other sectors include emissions from the food chain, including transportation, waste, and manufacturing.

For example, 3.6 percent of global greenhouse gas emissions come from waste, including landfills, wastewater, and other waste.[45] The food production system contributes its share to this total. After all, where does most of our uneaten food and food ready for harvest that never even makes it to our plates end up? Landfills. Solid waste, including food scraps, produces greenhouse gas emissions from anaerobic decomposition, which produces methane, and from carbon dioxide as a by-product of incineration and waste transportation.[46]

An additional 13.1 percent of the emissions that contribute to the global warming effect come from transportation, toting everything from people to pork chops.[47] The factory farming industry, in particular, demands energy-intensive shipping. CAFOs, for example, transport feed and live animals to feedlots and then to slaughter. Then the meat must be shipped to retail distribution centers and to the stores where it is sold to us consumers.

Americans, in particular, import and export a lot of meat. In 2007, the United States exported one 1.4 billion pounds of beef and veal (5.4 percent of our total production of beef)[48] and imported 3.1 billion pounds of the same.[49] One could argue that a lot of that transport is unnecessary from a consumer point of view and damaging from an environmental point of view.

Globally, international trade in meat is rapidly accelerating. As recently as 1995, Brazil was exporting less than half-a-million dollars' worth of beef. A little more than a decade later, the Brazilian Beef Industry and Exporters Association estimates the value of beef exports could reach $5.2 billion and expects revenues of $15 billion from beef exports by 2013.[50]

All of these billions of pounds of meat being shipped around the world add significantly to the carbon emissions from transportation. So do the Chilean grapes shipped to California, the Australian dairy destined for Japan, or the Twinkies toted across the country—all the meat and dairy, drinks, and processed foods shipped worldwide in today's globalized food market.

## THE ORGANIC SOLUTION

The globalized and industrialized food system has not only negative health consequences—think of all those Twinkies, that factory-farmed meat, and that chemically raised produce—but a climate change toll as well. But the news is not all bad. Once we gaze directly at the connection between food, farming, and global warming, we see plenty of cause for hope.

First, unlike many other climate change conundrums, we already know many of the steps we can take now to reduce carbon emissions from the food sector. For instance, we know that compared with industrial farms, small-scale organic and sustainable farms can significantly reduce the sector's emissions. Small-scale sustainable agriculture relies on people power, not heavy machinery, and depends on working with biological methods, not human-made chemicals, to increase soil fertility and handle pests. As a result, small-scale sustainable farms use much fewer fossil fuels and have been found to emit between one-half and two-thirds less carbon dioxide for every acre of production.[51]

We also are just beginning to see results from long-term studies showing how organic farms create healthy soil, which has greater capacity to store carbon, creating those all-important "carbon sinks."[52] By one estimate, converting 10,000 medium-sized farms to organic would store as much carbon in the soil as we would save in emissions if we took one million cars off the road.[53]

We're closer than ever to global consensus about the direction in which we need to head. In April 2008, a report on agriculture initiated by the World Bank, in partnership with the United Nations and representatives from the private sector, NGOs, and scientific institutions from around the world, declared that diverse, small-holder sustainable agriculture can play a vital role in reducing the environment impacts of the agriculture sector.

The result of four years of work by hundreds of scientists and reviewers,[54] the International Assessment of Agricultural Science and Technology for Development (IAASTD) calls for supporting agroecological systems; enhancing agricultural biodiversity; promoting small-scale farms; and encouraging the sustainable management of livestock, forest, and fisheries, as well as supporting "biological substitutes for agrochemicals" and "reducing the dependency of the agricultural sector on fossil fuels."[55] A civil society statement timed with the report's release declared that the IAASTD represents the beginning of a "new era of agriculture" and offers "a sobering account of the failure of industrial farming."[56] Said Greenpeace, the IAASTD report recommends a "significant departure from the destructive chemical-dependent, one-size-fits-all model of industrial agriculture."[57]

(Not everyone involved in the process was happy with the final report, which was signed by fifty-seven governments.[58] Chemical giant and agricultural biotechnology leaders Syngenta and Monsanto, for instance, refused to sign on to the final document. No public statements were given at the time.[59] But in an interview, Syngenta's Martin Clough told me, "When it became pretty evident that the breadth of technologies were not getting equal airtime, then I think the view was that there was no point in participating. It's important to represent the technological options and it's equally important to say that they get fair play. That wasn't happening."[60])

Despite the chemical industry holdouts, there is also consensus that sustainable farming practices create more resilient farms, better able to withstand the weather extremes of drought and flooding already afflicting many regions as a result of climate change. In other words, mitigation *is* adaptation. Because organic farms, by their design, build healthy soil, organic soils are better able to absorb water, making them more stable during floods, droughts, and extreme weather changes. In one specific example, conventional rice farmers in a region in Japan were nearly wiped out by an unusually cold summer, while organic farmers in the same region still yielded sixty to eighty percent of their typical production levels.[61]

In ongoing studies by the Pennsylvania-based Rodale Institute, organic crops outperformed nonorganic crops in times of drought, yielding thirty-five to one hundred percent more in drought years than conventional crops.[62] Visiting a Wisconsin organic farmer just after the major Midwest flooding of the summer of 2008, I could see the deep ravines in the surrounding corn fields caused by the recent flooding, while I spent the afternoon walking through a visibly unscathed biodiverse organic farm.

Encouraging sustainable agriculture will not only help us reduce emissions and adapt to the future climate chaos, it will have other beneficial ripples: addressing hunger and poverty, improving public health, and preserving biodiversity. In one study comparing organic and conventional agriculture in Europe, Canada, New Zealand, and the United States, researchers found that organic farming increased biodiversity at "every level of the food chain," from birds and mammals, to flora, all the way down to the bacteria in the soil.[63]

Finally, we know that shifting toward sustainable production need not mean sacrificing production. In one of the largest studies of sustainable agriculture, covering 286 projects in fifty-seven countries and including 12.6 million farmers, researchers from the University of Essex found a yield increase of seventy-nine percent when farmers shifted to sustainable farming across a wide variety of systems and crop types.[64] Harvests of some crops such as maize, potatoes, and beans increased one hundred percent.[65]

Here's the other great plus: we all have to eat, so we can each do our part to encourage the shift to organic, sustainable farming every time we make a choice about our food, from our local market, to our local restaurants, to our local food policies.

––––––––––

I was recently talking with Helene York, director of the Bon Appétit Management Company Foundation, an arm of the Bon Appétit catering company, which serves eighty million meals a year at four hundred venues across the country. York has been at the forefront of educating consumers and chefs about the impacts of our culinary choices on climate change, including leading the charge of the foundation's "Low Carbon Diet," which has dramatically reduced greenhouse gas emissions associated with their food. She summed up the challenge of awakening people to the food and climate change connection this way: "When you're sitting in front of a steaming plate of macaroni and cheese, you're not imagining plumes of greenhouse gases. You're thinking, dinner."

But the truth is those plumes of gases are there nonetheless, in the background of how our dinners are produced, processed, and shipped to our plates. Thankfully, more and more of us eaters and policymakers are considering the climate crisis at the end of our fork and what we can do to support the organic, local, sustainable food production that's better for the planet, more pleasing to the palette, and healthier for people too.

# Another Take | GLOBAL WARMING AND YOUR FOOD

## By the Cool Foods Campaign

The Cool Foods Campaign is a project of the Center for Food Safety and the Corner-Stone Campaign. Making the connections between the foods we eat and their contribution to global warming, the campaign aims to educate the public about the impact of its food choices across the entire food system and empower it with the resources to reduce this impact, focusing on agricultural practices and food choices that can reduce and reverse this trend. To learn more, visit Cool Foods' website at http://coolfoods campaign.org.

In "Global Warming and Your Food," the Cool Foods Campaign provides some additional information and ideas about the connections between the food we eat and the long-range future of our planet.

## IS INDUSTRIAL AGRICULTURE COOKING THE PLANET?

Did you know that our food system is a major contributor to global warming? The U.S. food system uses between seventeen and nineteen percent of the total energy supply in the country, contributing a significant amount of greenhouse gas emissions to the atmosphere every day.[1]

How is this possible? Greenhouse gases are generated in many ways, and many are created even before our food is grown.

On large-scale, modernized industrial farms (which traditionally grow only one or two crops—called monoculture—that rely heavily on pesticides, fertilizers, and fossil fuels), greenhouse gases are created in a multitude of ways. Pesticide and fertilizer applications, irrigation, lighting, transportation, and machinery are powered by greenhouse gas–emitting fossil fuels. The production of synthetic fertilizers and pesticides alone requires the equivalent use of more than 123 million barrels of oil, making them one of the largest contributors to greenhouse gas emissions in agriculture.[2]

The overuse of agricultural chemicals pollutes watersheds and kills plants that could otherwise capture greenhouse gases and actually reduce global warming. As the plants decompose, they emit methane, a greenhouse gas, into the atmosphere.[3] Methane is also emitted by the ninety-five million cows raised each year in the United States. The waste from these animals, and the sixty million hogs raised every year, are collected and stored in stagnant manure pits, which release not only a pungent smell but also more methane.

Once our food is grown, it is transported throughout the country to grocery stores and markets. The average American meal has traveled about 1,500 miles before it arrives on your plate. All told, the U.S food system uses the equivalent of more than 450 billion gallons of oil every year.[4]

## WHAT YOU CAN DO:
## REDUCING YOUR CARBON FOOTPRINT

You can have a major influence on global warming by making better food choices and reducing your "FoodPrint."[5] Your FoodPrint reflects the amount of greenhouse gases that were created in the production and shipping of the food that you buy. The "coolest" foods have the lowest FoodPrint and are made without producing excess greenhouse gases. When foods that produce higher FoodPrints—those considered "hot"—are avoided, we reduce our individual contributions to global warming. An easy way to tell if your food is "cool" or "hot" is to ask yourself these five basic questions before you buy.

- **Is this food organic?**
  Organic foods are produced without the use of energy-intensive synthetic pesticides and fertilizers, growth hormones, and antibiotics, and they are not genetically engineered or irradiated.

In addition to the emissions from producing fertilizers, nitrous oxide—a very potent greenhouse gas—is emitted when these chemicals are applied to farmland.[6] Conventional fertilizers also pollute water sources, which kill fish and plants and emit methane, also a very potent greenhouse gas.[7]

Unlike organic farming, conventional agriculture contributes to erosion by overusing synthetic pesticides. Not only does erosion emit carbon dioxide, but it transports agricultural chemicals to water sources.[8]

To Be Cooler: Buy organic and look for the USDA organic label to ensure that the food you eat is "certified organic."

- **Is this product made from an animal?**

  Conventional meats—e.g., beef, poultry, pork, dairy, and farmed seafood—are the number-one cause of global warming in our food system. Animals in industrial systems are fed foods they cannot biologically process. They are confined to unhealthy and overcrowded cages—conditions that contribute to malnutrition and disease. In an attempt to keep animals healthy, they are sprayed with more than two million pounds of insecticides, and their cages are sprayed with more than 360,000 pounds of insecticides every year.[9] They also ingest an astounding eighty-four percent of all the antimicrobials, including antibiotics, used annually in the United States.[10]

  Every year, livestock consume about half of all of the grains and oilseeds that are grown in the United States, thereby consuming more than fourteen billion pounds of fertilizers and more than 174 million pounds of pesticides. Producing all of these chemicals requires huge amounts of energy and is a major cause of global warming.

  To Be Cooler: Limit your consumption of conventional meat, dairy, and farmed seafood. Buy organic meat and dairy whenever possible, because these foods are produced without energy-intensive synthetic pesticides and herbicides, and look for wild (not farmed), local seafood.

- **Has this food been processed?**

  Compared to whole foods such as fruits and vegetables, processed foods require the use of energy-intensive processes such as freezing, canning, drying, and packaging. Processed foods are usually sold in packages that contain an ingredients label and are located in the center aisles of most grocery stores.

To Be Cooler: Do your best to avoid processed foods altogether, but "certified organic" processed foods are a good alternative.

- **How far did this food travel to reach my plate?**

  Transporting food throughout the world emits 30,800 tons of greenhouse gas every year. The average conventional food product travels about 1,500 miles to get to your grocery store.

  To Be Cooler: Choose locally produced foods or foods grown as close to your home as possible. Look for country of origin labels on whole foods and avoid products from far away.

- **Is this food excessively packaged?**

  Packaging materials, like plastic, are oil-based products that require energy to be created and are responsible for emitting 24,200 tons of greenhouse gas every year.

  To Be Cooler: Buy whole foods. Purchase loose fruits and vegetables (rather than bagged or shrink-wrapped); buy bulk beans, pasta, cereals, seeds, nuts, and grains; and carry your own reusable grocery bags.

## FOOD CHOICE AND BEYOND

Want to reduce global warming? Join our Cool Foods Campaign and help take a bite out of global warming by changing the way you eat.

The Cool Foods Campaign shows the connections between the foods we eat and their contribution to global warming. The aim of the campaign is to inform people about the impact of their food choices across the entire food system. We hope to inspire a groundswell of informed people committed to making sustainable food choices to reduce their FoodPrint. Our campaign seeks solutions to the problem of global warming and focuses on agricultural practices that can reduce and reverse this trend.

You can reduce your FoodPrint by making conscious food choices that contribute to the reduction in global warming. To keep up-to-date on the Cool Foods Campaign, and for more information about what you can do to lower your FoodPrint, keep visiting our website!

# SEVEN

# CHEAP FOOD
## *WORKERS PAY THE PRICE*

*By Arturo Rodriguez, with Alexa Delwiche
and Sheheryar Kaoosji*

A native of Texas, United Farm Workers (UFW) President Arturo S. Rodriguez has worked tirelessly to continue the legacy of Cesar Chavez since taking over the helm of the UFW upon the death of its legendary founder in 1993. The veteran farm labor organizer was first exposed to Chavez through a parish priest in his hometown of San Antonio in 1966. He became active with the UFW's grape boycott as a student at St. Mary's College in 1969. At the University of Michigan in 1971, where he earned a master's degree in social work, Rodriguez organized support for farmworker boycotts. Rodriguez has more than thirty-five years' experience organizing farmworkers, negotiating UFW contracts, and leading numerous farmworker boycott and political drives across North America.

Union membership has grown since Rodriguez kicked off an aggressive UFW field-organizing and contract-negotiating campaign in 1994. The union has made organizing a top priority, winning numerous elections and signing dozens of contracts with employers.

Among recent union victories under Rodriguez's leadership are agreements with D'Arrigo Bros., one of California's largest vegetable growers; Gallo Vineyards, Inc., America's biggest winery; Coastal Berry Co., the largest strawberry employer in the United States; Jackson & Perkins, the top rose producer in the nation; and Threemile Canyon Farms in eastern Oregon, the biggest dairy in the United States. Rodriguez has also established pacts protecting winery workers in Washington state and mushroom workers in Florida.

Alexa Delwiche joined the UFW as a researcher in 2006 after receiving her master's degree from the University of California at Los Angeles's School of Public Policy. Prior to graduate school, she was a research intern with the Institute for Food and Development Policy (Food First), where she researched the food crisis in Southern Africa as well as analyzed the relationship between food insecurity and childhood obesity in the United States. A focus of her work at UFW was to analyze food policy issues, which affect agricultural employers directly and subsequently impact farmworkers (such as international trade agreements, compliance with pending climate change regulations, and food safety), and to propose ways in which the UFW could promote forward-thinking, sustainable policies for

123

multiple stakeholders. Delwiche graduated summa cum laude and Phi Beta Kappa from the University of California at Santa Barbara in 2002. She left the UFW in late 2008 and now works for the San Francisco Board of Supervisors Office in the Office of the Legislative Analyst.

Sheheryar Kaoosji is a research analyst for the Change to Win (CtW) federation, a group of labor unions including the UFW, Teamsters, United Food and Commercial Workers, SEIU, Unite HERE, and Carpenters and Laborers. He has worked with several CtW unions on sectoral research, organizing strategy and policy advocacy. He has focused his career on developing models for sustainable economic development. Prior to joining the labor movement, he worked for community organizations working to retain working-class jobs and sustainable economic development in San Francisco. He holds a master's in public policy from the University of California at Los Angeles and a bachelor's degree from the University of California at Santa Cruz.

## DEATH IN THE FIELDS

Young grapevines thrive in the fierce summer sun of California's Central Valley. But the same early summer heat that helps bring life to the bountiful produce millions of Americans enjoy can also destroy. Unlike the young grapevines, assured of constant irrigation and hydration, farmworker Maria Isabel Vasquez Jimenez had to do without water as she labored in the fields in direct sunlight on a 95-degree day in May 2008.

After almost nine hours of work, Maria became dizzy and collapsed to the ground. Her boyfriend Florentino Bautista ran to her, held her in his arms, and begged for help. The foreman walked over to them and stood over the couple, reassuring Bautista and telling him that "this happens all the time." Remedies devised by the foreman and supervisor ranged from applying rubbing alcohol to placing a wet bandanna on Maria's body. Finally, Maria's boyfriend and coworkers were allowed to take her to a clinic, though the foreman told them to lie about where she was working. It took almost two hours to get the young woman to the clinic. Immediately upon her arrival, the clinic called an ambulance for the hospital. By the time she arrived at the hospital, her body temperature was 108.4 degrees.

Maria held on for two days, but her young body could not withstand the stress. Having arrived in the United States from a small village in Oaxaca only months before and having worked in the fields for only three days, seventeen-year-old Maria died, leaving her family and loving boyfriend forever. It's difficult to accept that such an injustice could occur in 2008. But tragic stories like Maria's are all too common in American agriculture.

Statistics tell part of the story. The rate of death due to heat stress for farmworkers is twenty times greater than for the general population.[1] In the past five years, thirty-four farm workers have died due to heat exposure in the United States. Six of those deaths occurred in the summer of 2008 alone.[2] The actual number is likely much higher because many farmworker deaths are not recorded as heat deaths, and some are not recorded at all.

Maria's death is a poignant example of how the pressures for decreased prices in our food system inevitably lead to exploitation of the workers at the lowest end of the economic chain. Food producers and retailers have successfully abdicated responsibility for the well-being of the workforce that makes their profits possible, aided and abetted by a callous and uninvolved government.

Who is to blame for these senseless and preventable deaths? The farm labor contractors who are hired by the growers to provide them with workers? The growers who actually own the crops and employ the contractors? Or the State of California, which is tasked with enforcing labor regulations that mandate shade, drinking water, and rest breaks for farmworkers toiling in the sun-baked fields?

From a moral perspective, there is more than enough blame to go around. But the system of agricultural production our society has created is designed to shield major corporations from any legal responsibility for their actions or inactions. The layers of subcontracting built into American agriculture are designed to shift responsibility downward from the largest firms to the smallest. Charles Shaw wine, sold exclusively at Trader Joe's and nicknamed affectionately by its customers "Two Buck Chuck," is solely produced by Bronco Wine Company, the largest wine-grape grower in the United States.[3] West Coast Grape Farming, a subsidiary of Bronco Wine and owner of the vineyard in which Maria died, hired Merced Farm Labor, an independent farm labor contractor, to provide workers to harvest their grapes. While Maria and her boyfriend were not picking grapes directly for the Charles Shaw label, both Bronco Wine

Company and West Coast Grape Farming are owned and operated by the same Franzia family, which singlehandedly supplies more than 360 million bottles of wine to Trader Joe's each year.

If this supply chain seems unnecessarily complicated, well, that's the point. Maria Isabel's employer was neither Trader Joe's, nor Charles Shaw–producer and –winemaker Bronco Wines, but a farm labor contractor with no discernible assets and no traceable relationship to the product sold. Thus, the retailer and producer are shielded from responsibility. Maria Isabel may die, but Charles Shaw wine is still on sale for $1.99 per bottle. Trader Joe's liberal, consumer-friendly reputation is preserved. And a farm labor contractor in Merced, California, quietly goes out of business. But the system that led to Maria Isabel's death continues without change.

The precariousness of a farmworker's life extends far beyond heat deaths. Fatality and injury rates for farmwork rank second in the nation, second only to coal mining.[4] The U.S. Environmental Protection Agency (EPA) estimates that U.S. agricultural workers experience 10,000–20,000 acute pesticide-related illnesses each year, though they also admit that this is likely a significant underestimate.[5] Drinking water and sanitary conditions—basic rights most American workers take for granted—are denied to farmworkers on a daily basis.

The plight of many farmworkers is made worse by their tenuous legal status as U.S. residents. Guest workers brought in under the government-sponsored H2-A visa program are routinely cheated out of wages, forced to pay exorbitant fees to recruiters, and virtually held captive by employers who seize their documents. Slavery—not in a metaphorical or symbolic sense, but in the literal meaning of the word—still exists on farms scattered throughout the country. In 2004, the United States Department of Justice investigated 125 cases of slave labor on American farms, involving thousands of workers.[6] Their shaky legal status helps explain farmworkers' vulnerability. Workers hesitate to report labor law violations for fear of losing their jobs or being deported. But workplace fear is just one variable in a systemic problem: the institutionalized acceptance of farmworkers as second-class citizens. The injustices farmworkers experience are by no means accidental.

Most Americans would be horrified to realize that the foods they eat are produced under conditions like these. Their lack of knowledge about these realities is attributable not to public apathy but to deliberate obfuscation by the companies that market foods and unconscionable neglect by the government agencies that should be safeguarding workers. As food production has become increasingly complex, it has become nearly impossible for consumers to gather information on their food purchases. Product labels reveal little information about a food's origin and contents and tell consumers nothing of the plight of those harvesting and processing their food. As a result, farmworkers in America are left to suffer incredible poverty and abuse in an industry characterized by great wealth and enormous profits.

## RETAILER POWER AND ITS IMPACT ON FARMERS

Food production in the United States is a high-technology, modern process. Produce and animal varieties are meticulously bred, designed, and genetically engineered by food scientists, supported by large academic programs at public land-grant universities across the nation and in government agencies like the USDA and FDA. Products are packaged, distributed, and sold throughout the nation and exported worldwide using the same state-of-the-art, just-in-time logistics systems that move other goods throughout the international economy.

But despite the high-tech systems that streamline and control the flow of goods to the nation's dinner tables, a farmworker's job is remarkably similar to the life of Chinese and Japanese immigrants at the turn of the twentieth century, of Mexican migrants in the 1920s, of white Dust Bowl migrants during the Depression, or black and Mexican migrant workers since the 1940s. Crops eaten raw, such as grapes, strawberries, and lettuce, require careful and steady human hands to retain the physical and visual perfection required by the modern consumer. Much of the so-called specialty crops (that is, fruits and vegetables) that the nation consumes are grown in the heat of California's Central and Coachella valleys, where daytime temperatures stay in the triple digits throughout harvesting season.

Strawberry and lettuce workers bend thousands of times a day, picking valuable produce from the rich earth of the central coast of California that John Steinbeck once wrote about. Apple workers fill bags with up to seventy pounds of produce, carrying them up and down ladders across the Yakima Valley of Washington state. The story is retold every generation, with workers toiling for minimum wages to feed the population of the United States.

Most people are surprised to learn that the conditions of farm laborers have not been dramatically improved since the days of *Grapes of Wrath*. In fact, the story has gotten worse in recent years. The growers that employ farmworkers have been experiencing a historic squeeze in prices from the retailers that purchase their produce. The retail food sector has been inexorably consolidating for decades, culminating in the dominance of Wal-Mart over the grocery industry, representing between twenty and forty percent of the sales volume of various food products in the United States. Today, the top five retailers control over sixty percent of the market.[7] Thanks to this consolidation, a relative handful of companies wield enormous power over food suppliers—power they use to demand ever-lower prices for the fruits, vegetables, and other commodities they stock. Growers have responded to these pressures, providing picture-perfect produce at deflationary prices, sourcing them from around the world in order to provide year-round supplies to the retailers.

And consumers have reaped the benefits. The average American family now spends less than ten percent of its income on food, the lowest percent in history.[8] In 1950, this figure was twenty percent. As writer and grower David Mas Masumoto described it in a poem he presented at the 2008 Slow Food Festival, "I remember $2-a-box peaches in 1961 and $2-a-box peaches in 2007."[9]

But everyday low prices for consumers (and increasing profits for the shareholders of the giant retailers) have created serious consequences for growers. Each year, farmers capture less of the consumers' food dollar. For example, in 1982, farmers received thirty-four percent and thirty-three percent of what consumers paid for fresh vegetables and fresh fruit, respectively, at retail food stores; by 2004, these farm shares had declined to nineteen percent for fresh vegetables and twenty percent for fresh fruit.[10]

One result of this squeeze is that only gigantic agribusinesses can survive in the new farming economy. Massive state-sponsored infrastructure, especially the dam and levee system developed in California over the twentieth century, allowed large California growers to achieve unprecedented scale while successfully gobbling up smaller growers farm by farm. Such scale also benefits top retailers that depend on the considerable quantities supplied by only the largest growers. Medium-sized family growers now depend on selling prime agricultural land to the developers who are helping California cities sprawl to supplement their income. Organic growers who once hoped to provide a new model of agriculture have been pushed to the periphery by high certification costs and loose standards that allow the largest growers to dominate the mainstream organic market.

But ultimately the effects of retail consolidation have trickled down the production line to workers, causing wages and benefits for workers throughout the food system to stagnate. Today, even a modest wage increase for workers on a farm could threaten a supplier's contract with a retailer. Furthermore, growers and labor contractors often find the easiest way to minimize labor costs is by cutting corners on labor and safety standards. Some labor contractors even admit that "breaking the law is the only way you can make decent money."[11] Of course, there are exceptions, and many decent employers in the industry exist. But the system results in a constant push for lower costs, with no basic standards across the industry. There is no high road—only a low road populated by growers racing one another to reach rock bottom.

## AGRICULTURAL "EXCEPTIONALISM" AND THE SUBSIDY ECONOMY

This situation is not sustainable, nor is it accidental. In large measure, it can be traced back to government policies designed to produce the very system that now distorts agricultural production in this country.

The State of California, where almost half of U.S. produce is grown and forty percent of farm laborers work, provides a vivid illustration of how this process has worked. The policies pursued by California as well

as by the federal government have promoted corporate agricultural interests for over a century.[12] By demanding a set of immigration policies
that guaranteed cheap labor and made these workers dependent on the
government through assistance programs (or on the growers themselves
through labor camps), a permanent underclass in the fields was cultivated.[13] This is a policy called "agricultural exceptionalism"—the ex-
emption of agriculture from labor and other laws under the Jeffersonian
theory that food production is not only a crucial U.S. industry but also a
superior way of life that deserves special preservation and protection.
This mindset has dominated U.S. policy in land use, labor law, and direct and indirect farm subsidies for a century.

As a result, although government exists in part to protect the powerless from the powerful, government has done little to benefit the agricultural workers who feed the nation. While most workers won historic
labor protections through the Fair Labor Standards Act of 1938—which
set minimum wage requirements, overtime laws, and child labor laws—
powerful lobbying by agribusiness succeeded in excluding farmworkers
from protection under these laws.[14] The minimum wage now applies to
farmworkers in most cases, but overtime provisions still do not. Furthermore, to this day, the age limit for children working in agriculture
differs from other businesses. Children under the age of 14 cannot be
employed in any other industry; in agriculture the age limit is twelve. No
age restrictions apply to children working on family farms.

Agribusiness also managed to exclude farmworkers from another
piece of New Deal legislation, the National Labor Relations Act, which
gave workers the right to form unions and bargain collectively. So when
the limited rights farmworkers actually enjoy are violated, they are denied the ability to organize themselves to demand fair treatment.

One of the most egregious examples of agricultural exceptionalism has
been the government's failure to reduce farmworkers' exposure to pesticides because of the profit loss growers might suffer if pesticides were
more tightly regulated. Cleverly, regulation of farmworker exposure to
pesticides was placed under the jurisdiction of the EPA, famous for its use
of "cost-benefit analysis" when determining whether to place restrictions
on the use of chemicals.[15] The phrase sounds innocuous, even reasonable,

but, in practice, reliance on cost-benefit analysis means that a hazardous pesticide will not be restricted by the EPA if the economic hardship to the grower is considered to be greater than the hazards to farmworker or consumer health. If pesticide protection for farmworkers were under the jurisdiction of the Occupational Safety and Health Administration (OSHA), cost-benefit analysis would not be required—a simple finding that workers' lives were at risk would suffice to justify regulation.

More than twenty years ago, the EPA concluded that farmworkers were disproportionately affected by the use of pesticides. Indeed, a large body of scientific literature has documented this relationship. A study of 146,000 California Hispanic farmworkers concluded that, when compared to the general Hispanic population, farmworkers were more likely to develop certain types of leukemia by fifty-nine percent, stomach cancer by seventy percent, cervical cancer by sixty-three percent, and uterine cancer by sixty-eight percent.[16] Despite this evidence, the EPA has failed to enact any mitigation procedures to address farmworkers' chronic exposure to pesticides.

Thanks to the importance of agriculture to the California economy, agribusiness has been especially powerful at the state government level. One result has been an amazingly lopsided set of water management policies: only three rivers in California remain undammed, and eighty percent of the water collected is consumed by agriculture.[17] Both small- and large-scale growers purchase vast amounts of imported river water at just a fraction of the true price, their purchases subsidized by state and federal taxpayer dollars. Indeed, in some regions of California, the average urban water user pays seventy-five times the price of a grower. Not only have these subsidies encouraged extremely inefficient water use by agricultural producers, but they have allowed growers and retailers to profit in their ability to produce crops more cheaply than otherwise possible, all at the expense of the taxpayer. Meanwhile, more than 635 miles of rivers and streams in the Central Valley have been classified as unsafe for fishing, swimming, or drinking due to pollution from agricultural runoff.[18] And as drought risk intensifies throughout the state, the bulk of the state's water continues to be dumped, quite literally, into the Central Valley for growers either to use for crops or to resell for profit to urban water districts, desperate for the precious resource.

## IMMIGRATION POLICY AS
## A TOOL OF AGRIBUSINESS

Over the last century, the most important policy perpetuating the existence of an underclass of agricultural workers has been U.S. immigration policy. Today, almost eighty percent of the 2.5 million farm workers in the United States were born outside of the United States.[19] The overwhelming majority of farmworkers are from Mexico. Estimates vary, but at least fifty percent of the workforce is not authorized to work in the United States.[20]

The roots of Mexican migration patterns lie in a series of U.S.-approved guest programs created during the twentieth century specifically to address farm labor needs and in lax enforcement of immigration laws by the federal government during periods in which no guest worker programs were in place.[21]

The most recent wave of labor migration to the United States was spurred by the North American Free Trade Agreement (NAFTA) and other trade liberalizing policies. The entrance of cheap, government-subsidized U.S. corn into the Mexican economy in the wake of NAFTA signaled to Mexican corn producers that there was no future left for them in agriculture. And they were right. Government investment in Mexican agriculture fell by ninety percent.[22] Almost two million subsistence farmers were displaced. Some farmers fled to urban areas in Mexico, while others decided there was nowhere else to go but north of the border.[23]

So NAFTA, after having been branded by both Republicans and Democrats as the solution to combating poverty in Mexico and reducing migrations to the United States, only deepened the problem. Rather than improving the economic situation on both sides of the border as promised, NAFTA helped produce a *decrease* in real wages in Mexico between 1995 and 2005.[24] Annual migration to the United States *increased* from 2.5 million unauthorized immigrants in 1995 to 11 million in 2005.[25] As a result of the effects of NAFTA on food production in southern Mexico, Oaxacans are now the fastest growing population of farmworkers in the United States. Today, one in five families in Mexico depends on remittances from the United States, the total averaging almost $24 billion a year.[26]

Since NAFTA's implementation, the only serious U.S. immigration policy enacted to address the influx of migrants was Operation Gatekeeper (1995), the policy of deterring migration by increasing border enforcement in the border cities of San Diego and El Paso.[27] This policy only shifted migration patterns farther east into more harsh and inhospitable desert. Mexican farmworkers leave their families and risk their lives crossing a dangerous and increasingly militarized border because there are no other options left. To risk is to hope for better.

Tragically, since 1995, deaths along the United States/Mexico border have doubled.[28] In 2005, 472 people died in the desert, with heat exposure and dehydration as the leading causes of death. And when migrants do survive, their best and often only shot at employment in the United States is the lowest-paid, most-dangerous, least-respected occupations of all—including agriculture. In a world of free trade for goods and closed borders for people, those who survive the passage can check their rights at the gate.

## FARM WORKERS LOOKING UP
## AT THE POVERTY LINE

It's difficult for a family of four living at the official poverty level to get by, making just $21,000 per year.[29] But compare that with a farmworker family's annual wage of roughly $13,000, a sum comparable to one acre's profit for a strawberry field.[30] Farmworkers continue to be among the lowest-paid laborers in the United States, with only dishwashers earning less.[31] The majority of farmworkers live below the poverty line, with real wages hovering around minimum wage.[32] Benefits are even more meager. Less than one-tenth of workers have employer-paid health insurance for non–work related health care, and just ten percent receive paid holidays or vacation time.[33]

As we've noted, a significant factor accounting for the low wages of farmworkers is the large proportion of unauthorized immigrant workers employed in agriculture. Undocumented workers have few employment options, thus making workers more willing to accept low wages.[34] The use of farm labor contractors (FLCs), hired by growers to provide workers, also helps explain the low wages for farmworkers. FLCs are notorious for paying lower wages in order to compete with the other thousands of contractors desperate for a grower's business. Farm labor contractors provide almost fifty percent of farm labor in California, a number that grows every year both in California and in the rest of the nation.

U.S. policymakers consistently support and facilitate agricultural producers' dependence on low-wage labor. Some may believe that the poor compensation of immigrant workers will be augmented by a safety net of government assistance programs designed to help the working poor. Yet undocumented status and extreme poverty have proven to be insurmountable barriers that prevent many farmworker families from actually accessing much of this assistance.

While the legal status of workers is no impediment when hiring workers for poverty-level wages, the same status excludes these workers from their only opportunities to make ends meet through government assistance. Less than one in five farmworkers use means-tested services such as Temporary Assistance to Needy Families; housing vouchers; Women, Infants, and Children; food stamps; Medicaid; or the National School Lunch Program, citing legal status and cost as significant barriers.[35] And even though each week a percentage of farmworkers' paychecks flows into the Social Security system, most will never see a dime of it. Only two percent of farmworkers report ever receiving any social security benefits.[36] Far from being a drain to the system, those working without authorization actually provide Social Security with an average annual subsidy of nearly $7 billion.[37]

High rates of poverty contribute to shocking health problems among farmworkers. Nearly eighty percent of male farmworkers are overweight.[38] And one in five males have at least two of three risk factors for chronic disease, such as high cholesterol, high blood pressure, or obesity, putting them at heightened risk to suffer from heart disease, stroke, asthma, and diabetes. Many farmworkers have never visited a doctor's office or any other type of medical facility, including an emergency room. And as migrant workers return to Mexico, the costs associated with medical treatment for their high rates of chronic disease will be borne by Mexican society, thus creating a greater strain on the already fragile Mexican economy.

Poverty rates among farmworkers extend into their housing options. Interestingly, in 2005, the three least affordable places to live in the

United States, measured by the percentage of income spent on rent or mortgage payments, were areas with high farmworker populations: Salinas, CA; Watsonville, CA; and Petaluma/Santa Rosa, CA.[39] In expensive cities such as these, affordable housing options are in such short supply that workers often are forced to live wherever they can find shelter, from abandoned cars to tin-roofed shanties.

Squalid living conditions in labor camps, reminiscent of those we associate with the 1930s, still exist for farmworkers today. In May 2008, more than one hundred migrant fruit pickers from Washington were found living in tents in a Central Valley cherry orchard in California without access to clean water. They were bathing in drainage ditches.

There have been some model housing programs created in California and Florida, but generally speaking, substandard living conditions are the norm for farmworkers. While some workers live in employer-provided housing, many families instead choose to crowd into rented apartments, sharing space appropriate for a single family with as many as ten to twenty other people. Poor sanitation and proximity to pesticide-laden fields create serious public health risks for farmworker families. But these are the conditions workers must accept when left with no other options.

## GOVERNMENT'S MALIGN NEGLECT

It isn't enough to demand that government should create more laws to protect farmworkers. The ones that do exist aren't enforced. Agricultural land is too vast for government to patrol; there are more than 80,000 farms in California alone, employing well over half a million workers.[40] CAL-OSHA, the state agency responsible for worker safety enforcement, conducted fewer than 300 inspections between 2007 and 2008.[41] Penalties are meager, and often violations are never even collected.

California holds the reputation for the most pro-farmworker legislation in the nation, yet the state provides many examples of the inadequacies of government enforcement. California remains the only state to enact any legislation granting farmworkers basic labor rights. The Agricultural Labor Relations Act (ALRA) of 1975 provides agricultural workers with organizing rights and protects workers from workplace retaliation from employers due to union involvement. However, even after unions gained organizing rights, the State of California ignored most of these rights once conservatives had regained the state house in 1983. The growers reasserted their ability to intimidate, and the United Farm Workers' ability to organize workers returned to the pre-law level that year, when the Republican-appointed Agricultural Labor Relations Board determined that even the cold-blooded murder of pro-union farmworker Rene Lopez by a company goon did not amount to an unfair labor practice.

Despite the existence of ALRA, union density rates among farmworkers remain low. Growers were found by the State of California to have intimidated, threatened, and offered bribes to workers at union elections in 2005 and 2006, but there have been no meaningful penalties, only an offer of another chance for workers to try to organize under such conditions.[42]

The failure to enforce heat protection laws provides another example. In 2005, California passed a law to protect workers from death due to heat stroke. The law requires that employers provide fresh water, shade, and additional breaks when the temperature goes above 95 degrees. Yet more farmworkers have died in the three years since the law's enactment than the three years prior.[43] More than one-third of the farms visited by CAL-OSHA in 2007 were out of compliance with the heat regulations.

Pesticide spraying provides yet another example. California leads the nation in pesticide protection for farmworkers, which includes one of only two pesticide spraying reporting systems in the nation. Even so, pesticide spraying continues at high rates, and the burden for reporting rests with the workers, keeping reporting rates low.

## THE UNION IS THE SOLUTION

Farmworkers have long known that, though government may try to protect them, ultimately it will not stand with them. Workers without representation have no chance of even bringing a claim to enforcement agencies without facing threats of firing, deportation, or worse. The most effective solution for workers has been to organize and collectively demand improved wages, benefits, and working conditions. Unionized workers have the protections that allow them to speak up when something goes wrong.

Unfortunately, the policy of excluding agriculture from laws that affect the rest of the country (especially labor law) lives on. And rather than a remnant of the past, it may in fact be a harbinger of where the American economy is headed in the twenty-first century. The abused and exploited second tier of the American labor force has expanded from agricultural workers to include many food-processing workers across the nation as well as employees in other sectors dominated by people of color, such as hospitality, janitorial work, trucking, and security. Millions of these workers effectively have no rights and no chance of basic protections without organizing themselves into unions.

The history of the meatpacking industry in the twentieth century is an example of the effect unions can have on an industry. Long one of the dirtiest, lowest-paying, most-dangerous urban jobs, as depicted in Upton Sinclair's 1906 novel *The Jungle*, meatpacking was transformed in the 1930s into a well-paying, respected, and safer job because the workers organized the Meatpackers' Union. The daily tasks of these jobs were not transformed by unionization, but the union allowed its members to improve their working conditions, raise families, buy homes, and even overcome the racial discrimination that existed in postwar cities such as Chicago and Minneapolis. At the same time, food safety and quality were boosted by workers with such crucial rights as whistle-blowing and the ability to affect assembly line speeds, benefiting not just workers but consumers.

In recent years, the same price pressures from retailers such as Wal-Mart that have depressed agricultural wages have resulted in meat processing being moved to rural, nonunion facilities run by such strongly antiunion meat and poultry firms as Tyson and Smithfield. The result has been declining standards, wages, and quality in the meatpacking industry. The jobs that once supported the working class in Midwestern cities have been redesigned to exploit workers from around the world, who have no ability to complain about or question unsafe and unfair practices. Progress has turned back, to the detriment of both rural and urban economies, workers, and consumers. Once again, only big retailers and food processors win.

There is hope for better conditions. Only about five percent of U.S. farmworkers are unionized. The unionized workers don't always make much more than other farmworkers (though some do), but they do get health insurance and even a pension, which are increasingly rare even for white-collar workers. More important, union workers have the right to speak up on the job. They have the freedom to advocate for issues that matter to them without fear of dismissal. They can establish procedures with their supervisors for days it gets too hot in the fields and workers feel their health may be at risk. The value of this is remarkable both for the basic conditions of the workplace and the quality and sustainability of the food produced. And a farmworker union with higher density would be a powerful ally to the environmental justice movement that is rising across the nation, empowered to address global as well as kitchen-table issues in the food processing industry.

Cesar Chavez, founder and leader of the United Farm Workers until his death in 1993, focused much of his work on environmental issues such as pesticide use, the dangers of monoculture, and the benefits of natural production methods, understanding that collective bargaining was a meaningless tool if farmworkers were still being poisoned in the fields. He recognized that pesticide use is an issue that clearly affects both workers and consumers.

The safety and well-being of farmworkers, consumers, and the environment have always been secondary to profits in the eyes of the big agricultural interests. So, when the government failed to address concerns about the harmful effects of organophosphates on workers, consumers, and the environment, Chavez and the UFW used union contracts to regulate pesticides. Union contracts in 1970 achieved what no U.S. government agency ever had: key provisions restricting the use of the five most dangerous pesticides.[44]

Unfortunately, the current system for farmworkers to choose a union is broken. The sad truth is that the federal agencies designed to oversee the food system and its workforce—the Food and Drug Administration, the U.S. Department of Agriculture, the Department of Labor, and the Environmental Protection Agency—as well as the State of California's Agricultural Labor Relations Board, have little practical ability to conduct a fair union election, let alone regulate pesticides, food safety, or worker safety in the fields.

It is even worse in other states. The agricultural exceptionalism of the 1930s remains dominant, and there is no way for farmworkers to organize in the rest of the country. Many young farmworkers don't know what a union is, having grown up in a post-NAFTA North America of economic instability, migration, and constant fear.

## FARM WORKERS AND CONSUMER POWER

This litany of problems has a solution. It is to change the balance of power when it comes to our food. Workers have never had the power to balance out the strength of agribusiness. But consumers have enormous power when they are activated and informed.

The two UFW-led grape boycotts of the 1960s and 1970s were unprecedented in their scope, duration, and effectiveness because they combined the power of the farmworker, on strike in California, and the consumer, refusing to purchase the product across the country. This boycott was able to defeat the power of the retailer and grower because the consumer, the final arbiter of the transaction, took action.

So why has the consumer abdicated this power in more recent years? Have consumers made a Faustian bargain, accepting worker exploitation in exchange for low prices? If so, the actual benefits to consumers are meager. University of California at Davis agricultural economist Philip Martin has computed that farmworker wages and benefits levy a total cost of $22 per year on each American household.[45] Furthermore, he found that to raise average farmworker wages by forty percent, bringing workers from below the poverty line to above it, would cost the average household only $8 more for produce each year.[46]

Thankfully, consumer awareness in the food system has reached unprecedented levels, and the ability to create change to the food system has made important strides because of consumer preference. The organics industry continues to grow. Wal-Mart, long criticized for its irresponsible buying practices and cheap products, now carries organics and purchases locally grown produce when possible. Concern over the treatment of animals has led to ballot propositions banning caged animals and has changed the buying practices of major fast-food outlets.

The growth explosion of Whole Foods from a niche natural foods store to a $6 billion powerhouse proves that consumers are willing to pay more for quality food.

The principle of sustainability recognizes the interdependence of our food system. And worker dignity, respect, and health and safety are fundamental to a sustainable system. Purchasing organic strawberries doesn't mean much if workers are still dying in fields. This same force that exploits farmworkers also pollutes our environment, impoverishes rural communities, and sickens consumers. Unless the balance of power is shifted away from valuing profits over human life, no one is protected.

Cesar Chavez once noted, "In the old days, miners would carry birds with them to warn against poison gas. Hopefully, the birds would die before the miners. Farmworkers are society's canaries."[47] The integrity of the food system begins with just conditions for workers.

Only the consumer has the power to support farmworkers in their struggle for representation. And only with an empowered workforce will there be an organized, principled counterbalance to the food production sector, defending sustainability, safety, and other standards. When workers are empowered, consumers are protected. Working together to take the following steps, we can make our food system more just, sustainable, and healthy.

## WHAT CAN YOU DO?

If you share the concerns we've described in these pages, here are some practical steps you can take to support our efforts to improve the conditions of the workers who provide you with the food that you and your family eat:

- Become educated on farmworker issues. Start by visiting the resource library at www.ufw.org.
- Support union and other advocacy campaigns for workers in the food system.
- Demand that retailers provide more transparent information on the working and living conditions of their suppliers' workforce.

- When they're available, always purchase products that guarantee workers' rights and express your support and approval to retailers so they'll be encouraged to stock such products.
- Support comprehensive immigration reform.
- Support policies that assist the working poor, such as increasing the minimum wage, living wage ordinances, and universal health care.
- Buy organic. Even though organic production does not provide workers with any additional wages, benefits, or respect, they are spared the detrimental effects associated with pesticides.

While it's time for consumers to mobilize and participate in reforming the food system, it is equally important for farmworker and food-system advocates from every step of the supply chain to come together and engage in a serious dialogue. It is our responsibility as advocates to create opportunities for consumers to use their power in improving the lives of food-system workers. Whether this occurs through the development of a "Socially Just" food certification label, through an extensive consumer awareness campaign, through legislation, or through an entirely new vision, now is the time to organize ourselves, to work together, to implement the type of societal transformation that we envision every morning when we get out of bed, and to renew our struggle for economic and social justice.

*By the Pesticide Action Network North America*

---

Pesticide Action Network North America (PANNA) works to replace pesticide use with ecologically sound and socially just alternatives. As one of five autonomous PAN regional centers worldwide, PANNA links local and international consumer, labor, health, environment, and agriculture groups into an international citizens' action network that challenges the global proliferation of pesticides, defends basic rights to health and environmental quality, and works to ensure the transition to a just and viable society. To learn more about PANNA's work, visit its website at http://www.panna.org.

In "Fields of Poison," PANNA explains the impact of pesticide use on the people most directly affected and most frequently harmed—the farmworkers who are forced to come into contact with these often-dangerous chemicals.

---

Agricultural workers face a greater threat of suffering from pesticide-related illness than any other sector of society. On a daily basis, their livelihood requires working with pesticides either by mixing and applying the chemicals directly or by planting, weeding, harvesting, and processing treated crops. Because farmworkers and their families often live in or near treated fields, pesticide exposure threats exceed even those understood to go along with pesticide application and mixing or handling treated crops. For instance, most reported incidences of poisoning happen because of airborne pesticide drift.

Pesticide-related illness ranges from cases of acute poisoning—with symptoms including nausea, dizziness, numbness, and death—to

143

pathologies whose origins are more difficult to trace, like cancer, developmental disorders, male infertility, and birth defects. Among the additional challenges faced by farmworkers are the lack of health care, legal representation, and often social standing required to make known the risks and costs that they and their families bear in order to put food on the table. Over the last twenty years, however, data around pesticide-poisonings in California's agricultural fields have slowly and imperfectly started to take shape as more farmworkers and farmworker advocates are finding venues to report their experiences.

California is among the only states to gather information on pesticide poisonings in agricultural workers, and until the last decade, that information was not easily accessible to the public. To shed light on the issue, the statewide coalition Californians for Pesticide Reform (CPR) published *Fields of Poison: California Farmworkers and Pesticides* in June 1999. Using government reports, worker testimonials, and other resources, *Fields of Poison* described myriad barriers to reporting pesticide-related illnesses and concluded that *reported illnesses represented only the tip of the iceberg of a serious and pervasive problem.* Pesticide Action Network (PAN) has worked with CPR, United Farm Workers of America (UFW), and California Rural Legal Assistance Foundation (CRLAF) to continue tracking pesticide poisonings in the fields of California in the decade since that original report.

Statewide, reported agricultural pesticide poisonings have decreased from a yearly average of 665 cases (1991–1996) to 475 (1997–2000) to 415 (2000–2006) (see Table 7.1). According to advocates and farmworkers in the field, however, many cases go unreported, so true figures are probably much higher. One especially sharp decline in the number of reported cases in 2006 is attributed by the Department of Pesticide Regulation (DPR) to inadequate funding for several government programs that facilitate reporting.[1] Other ongoing challenges to accurate reporting may include doctors' failure to recognize and/or report pesticide-related illnesses; failure of insurance companies to forward doctors' illness reports to the proper authorities; or farmworker reluctance to seek medical attention for suspected pesticide exposure for fear of losing their jobs.

After twenty years of monitoring the situation, PAN, CPR, UFW, and CRLAF have found that uneven reporting practices, lax worker protection standards, and regulatory enforcement models prove inadequate tools to protect farmworkers from pesticide poisoning in our agricultural fields. Only elimination of hazardous pesticides and their replacement with safer, less toxic pest management tools is a truly sustainable solution to agricultural chemical exposure. Persistent effort to reduce and eliminate use of hazardous pesticides through development and implementation of ecologically sustainable production methods is the cornerstone for reducing the burden of acute and chronic pesticide illness.

## PESTICIDES INVOLVED IN POISONING CASES ARE AMONG THE MOST HAZARDOUS

Over the period from 1998 to 2006, fourteen of the top twenty pesticides linked to reported illnesses are classified as particularly hazardous Bad Actors (see table). The fumigant metam-sodium was the most frequently listed Bad Actor. Also of particular note is the number of cases due to exposure to the organophosphate nerve toxin insecticides. For example, agriculture continues to widely use chlorpyrifos—banned for all home use in 2002.

## SOIL FUMIGATION LEADS "BAD ACTOR" LIST IN NUMBERS OF POISONINGS

Since soil was first identified as an application site in 1998, reported poisonings related to soil fumigations have been on the rise. (Fumigant pesticides—those applied as a gas or that quickly become gases—present a special case of exposure through the air. Fumigants are extremely hazardous substances used primarily to sterilize the soil before planting. They are the antithesis of ecological, sustainable agriculture and an increasingly common culprit in farmworker pesticide exposure.) While only 222 soil-related cases were reported from 1998 to 2000, 1,149 cases were recorded from 2001 to 2006, over three times the number reported from any other crop. All but one of these incidences involved a "Bad Actor" fumigant such as metam-sodium, methyl bromide, chloropicrin, metam-potassium, or 1,3 dichloropropene.

| Top 20 Pesticides Implicated in Reported Poisoning Cases, 1998–2006 | | | |
|---|---|---|---|
| Pesticide | Cases 98–06 | Bad Actor* | Hazard |
| Not determined | 1,360 | | |
| Metam-Sodium | 646 | Yes | Developmental toxin, carcinogen |
| Adjuvant | 604 | | |
| Sulfur | 453 | | |
| Chlorpyrifos | 239 | Yes | Nerve toxin, moderate acute toxicity, suspected endocrine disruptor |
| Glyphosphate | 146 | | |
| Methomyl | 119 | Yes | Nerve toxin, high acute toxicity, suspected endocrine disruptor |
| Methyl Bromide | 113 | Yes | High acute toxicity, developmental toxin |
| Sodium Hypochlorite | 110 | Yes | High acute toxicity |
| Dimethoate | 103 | Yes | Nerve toxin, high acute toxicity, developmental toxin, possible carcinogen |
| Spinosad | 100 | | |
| Propargite | 66 | Yes | High acute toxicity, developmental toxin, carcinogen |
| Petroleum Oil | 59 | | |
| Carbofuran | 40 | Yes | Nerve toxin, high acute toxicity |
| Diazinon | 38 | Yes | Nerve toxin, moderate acute toxicity, developmental toxin |
| Myclobutanil | 38 | Yes | Slight acute toxicity, developmental toxin |
| Naled | 36 | Yes | Nerve toxin, moderate acute toxicity, developmental toxin |
| Copper Hydroxide | 36 | | |
| Iprodione | 35 | Yes | Slight acute toxicity, carcinogen |
| Oxydemeton-Methyl | 32 | Yes | Nerve toxin, high acute toxicity, developmental toxin |
| Esfenvalerate | 28 | | |
| Mancozeb | 26 | Yes | Developmental toxin, carcinogen |

SOURCE: California DPR PISP data and the PAN online pesticide database (www.pesticideinfo.org). *PAN uses Bad Actor to describe pesticides that are 1) known or probable carcinogens; 2) reproductive or developmental toxicants; 3) neurotoxic cholinesterase inhibitors; 4) known groundwater contaminants; or 5) of high acute toxicity.

## WORKER SAFETY REGULATIONS
## ARE INADEQUATE AND OFTEN VIOLATED

Farmworkers have lacked basic protections enjoyed by workers in other industries for decades. When it comes to regulatory protection from workplace poisoning by pesticide exposure, the situation is no different. To take a representative sample: fifty-one percent of poisoning cases from 1998 to 2000 occurred when pesticides drifted from site of application onto workers. Another twenty-five percent resulted from dermal contact with pesticide residues. DPR found no relevant violations in 286 (forty-two percent) and 189 (fifty-six percent) of drift and residue cases, respectively. In other words, in a substantial number of cases, existing laws and regulations failed to protect workers from poisoning.

## DPR REPORTS REVEAL WIDESPREAD
## VIOLATIONS AND INVESTIGATION FLAWS

From 1997 to 2001, DPR staff observed 572 pesticide-related field operations in twenty counties and reported that over one-third violated one or more safety regulations. Common violations included failure to provide useable protective equipment, washing/decontamination facilities, and fieldworker access to pesticide use information. DPR found that eighty-eight percent of protective equipment violations were due to employer negligence and only twelve percent to worker failure to utilize available protective equipment.

A DPR review of county illness investigations revealed serious investigation flaws, including interviewing workers in the presence of their employers and using employer-affiliated translators at least one-third of the time. A DPR analysis of illness episodes between 1991 and 1999 showed that sixty-eight percent of early reentry illness episodes were due to failure to notify workers that a field was under a restricted entry interval. In the California Agricultural Workers' Health Survey conducted by an independent research institute, only fifty-seven percent of farmworkers surveyed in seven California communities reported receiving pesticide safety training that is required under the federal Worker Protection Standard.

## POOR ENFORCEMENT OF LAWS, MOST COUNTY AGRICULTURAL COMMISSIONERS STILL ISSUE FEW FINES

California county agricultural commissioners continue to issue few fines when violations are found, responding instead with letters of warning and violation notices. During fiscal year 2000/2001, DPR issued only 520 fines statewide for agricultural pesticide safety violations, along with 4,069 letters of warning or notices of violation. Most fines ranged from $151 to $400, an amount DPR designates for moderate violations that pose a reasonable possibility of creating a health or environmental hazard or for repeat record-keeping violations. The annual number of fines in the moderate and serious categories has remained relatively constant since *Fields of Poison*'s original publication in 1999, but the number of fines for minor violations has dropped.

## SUMMARY: URGENT NEED FOR BETTER WORKER PROTECTIONS AND SAFER AGRICULTURE

Better enforcement models do exist, and the time for more comprehensive regulatory interventions is upon us. DPR and county agricultural commissioners share responsibility for regulating agricultural pesticide use in California. DPR's evaluation of enforcement program weaknesses is a good first step, but progress toward more effective enforcement has been slow. *Now* is the time to move beyond studying the problem and to start acting.

Use of hazardous pesticides and inadequate regulations continue to seriously threaten California farmworker health and well-being. Only elimination of hazardous pesticides and their replacement with safer, less toxic pest management tools is a truly sustainable solution to agricultural chemical exposure. Persistent effort to reduce and eliminate use of hazardous pesticides through development and implementation of ecologically sustainable production methods—already demonstrated as viable by a burgeoning organic agriculture movement in California and around the world—is the cornerstone for reducing the burden of acute and chronic pesticide illness.

For more information, contact the Pesticide Action Network North America, www.panna.org.

## EIGHT

# THE FINANCIAL CRISIS
# AND WORLD HUNGER

### By Muhammad Yunus

Muhammad Yunus was born on June 28, 1940, in the village of Bathua, in Hathazari, Chittagong, the business center of what was then Eastern Bengal. His father was a small local goldsmith who encouraged his sons to seek higher education. But his biggest influence was his mother, Sufia Khatun, who always helped any poor person who knocked on their door.

In 1974, Professor Muhammad Yunus, a Bangladeshi economist from Chittagong University, met a woman who made bamboo stools and learned that she had to borrow from a moneylender the equivalent of 15 cents to buy bamboo for each stool made, leaving her just a penny profit margin. Had she been able to borrow at more advantageous rates, she would have been able to improve her economic condition and raise herself above the subsistence level.

Realizing that there must be something terribly wrong with the economics he was teaching, Yunus took matters into his own hands, and from his own pocket lent the equivalent of $27 to forty-two poor people. With this he launched a program to demonstrate that it was possible, through tiny loans, not only to help poor people survive, but also to create the spark of personal initiative and enterprise necessary to pull themselves out of poverty.

Yunus carried on giving out micro-loans, and in 1983 formed the Grameen Bank, meaning "village bank," founded on principles of trust and solidarity. In Bangladesh today, Grameen Bank has 2,520 branches, with 27,000 staff serving 7.51 million borrowers in all the villages of Bangladesh. Grameen Bank lends out over a billion U.S. dollars annually. Ninety-seven percent of the borrowers are women, and more than ninety-eight percent of the loans are paid back.

Today, Grameen methods are being applied in projects in almost all countries of the world, including the United States, Canada, France, Italy, Spain, the Netherlands, and Norway. For their efforts to alleviate poverty, Yunus and Grameen Bank were named corecipients of the 2006 Nobel Peace Prize.

In 2008, under the crushing collapse of the U.S. financial system, the global economy suffered a massive shock wave. Now giant financial institutions, along with major manufacturing firms like the automakers, are going bankrupt or being kept alive only with unprecedented government bailout packages. Stock markets all over the world are reporting on a daily basis how companies and individuals are together losing trillions of dollars.

Many reasons have been suggested for this historic economic collapse: excessive greed in the marketplace, the transformation of investment markets into gambling casinos, the failure of regulatory institutions, and so on. But one thing is clear: the rich will not be the worst sufferers from this financial crisis. Instead, it will be the bottom three billion people on this planet who will be harmed the most, despite the fact that they are not responsible in any way for creating this crisis. While the rich will continue to enjoy a privileged lifestyle, the bottom three billion people will face job and income losses that, for many, will make the difference between life and death.

The financial crisis is so shocking that world leaders and the media have almost forgotten that 2008 was also the year when another big crisis rocked the world—the food crisis. We have only seen the beginning of these crises in 2008; it is going to be a long and painful period ahead. The combined effects of the financial crisis and the food crisis will continue to unfold in the coming months and years, affecting the bottom three billion with special force.

During 2008, the United Nations World Food Programme (WFP) reported that more than seventy-three million people in seventy-eight countries were facing the reality of reduced food rations. We saw headlines reporting news of a sort many people assumed we would never experience again: skyrocketing prices for such staple foodstuffs as grains and vegetables (wheat alone having risen in price by 200 percent since the year 2000); food shortages in many countries; rising rates of death from malnutrition; even food riots threatening the stability of countries around the globe.

Since the latest peak in global food prices (which occurred in June 2008), prices have gone down, bringing a bit of short-term relief to millions. These high food prices have created tremendous pressure in the lives of poor people, for whom basic food can consume as much as two-thirds of their income. Sustained, generous, wise leadership and broad-based cooperation is required to overcome the continuing crisis and save the millions of still-threatened lives.

And while short-term relief efforts are essential to stave off the immediate effects of food shortages and prevent widespread famine, it's also important to step back and take a look at the broader causes of the crisis—to consider how the evolution of the world economy and, in particular, of the system whereby food is produced and distributed has led us to today's dilemma. Perhaps surprisingly, the economic, political, and business practices of the developed world, including the United States, have a profound impact on the availability of food in the poor nations of the world. Thus, solving the global food problem will require an international effort, not merely a series of local or even regional reforms.

Not so long ago, massive food shortages seemed to be a thing of the past. The Green Revolution of the 1950s and 1960s, spearheaded by the breakthrough scientific work of Dr. Norman Borlaug and other scientists, such as Indian agricultural researcher Monkombu Swaminathan, increased crop yields in Asia and Latin America and made many countries that had been reliant on food imports self-sufficient. Rates of hunger and malnutrition dropped significantly (although Africa, unfortunately, received little of the benefit because the local environmental needs of that continent were largely ignored in the research of the time). The high-yield grain production made possible by the Green Revolution has been credited with saving the lives of up to a billion people.

Now, however, a series of interrelated trends has partially reversed the gains that the Green Revolution produced.

Part of the problem has been the way in which the globalization of food markets has been managed over the past three decades. I am a strong proponent of free trade; I believe that encouraging people and nations to exchange goods and services with one another will, in the long run, lead to greater prosperity for all. But like all markets, global markets need reasonable rules that will allow all participants an opportunity to benefit. This is especially true when the market includes some players—such as the richest nations of North America, Europe, and Asia—who are highly advantaged, while others—such as the poor nations of the developing world—are struggling to gain a foothold.

Today's global markets, unfortunately, are only partly free, and some of the restrictions and distortions that have been left in place have had devastating consequences for poor nations. Since the 1980s, the developing countries have been pressured to open their markets to food imports from the developed world and to eliminate protected markets and subsidies for their own farmers. For example, agencies like the World Bank and the International Monetary Fund (IMF) have made loans to poor nations contingent on the requirement that recipient countries reduce governmental support for local food producers, and international arrangements like the North American Free Trade Agreement (NAFTA) have eliminated tariffs protecting local farmers. Yet at the same time, crop subsidies and export controls remain in place in many countries of the developed world, including the world's greatest food producer, the United States.

As *The Economist* has noted, these distortions have combined with a neglect of necessary research and development work to stall or reverse the agricultural advances that were formerly being made in the poor nations of the world:

> Most agricultural research in developing countries is financed by governments. In the 1980s, governments started to reduce green-revolutionary spending, either out of complacency (believing the problem of food had been licked), or because they preferred to involve the private sector. But many of the private firms brought in to replace state researchers turned out to be rent-seeking monopolists. And in the 1980s and 1990s huge farm surpluses from the rich world were being dumped on markets, depressing prices and returns on investment. Spending on farming as a share of total public spending in developing countries fell by half between 1980 and 2004. . . . This decline has had a slow, inevitable impact.[1]

The imbalances caused by this semi–free trade are distorting markets, raising prices, and even destroying agriculture in poor countries that once boasted enormous food surpluses.

Subsidies for ethanol in countries like the United States are one example of this problem. Intended to encourage the growth of corn and soy to partially replace fossil fuels in gasoline, these subsidies were designed to make it economically viable to use biofuels as a partial substitute for relatively cheap and abundant oil—and they worked as intended, as shown by the fact that, in 2007, fully one-quarter of the maize (corn) crop in the United States was used to manufacture ethanol.

Ethanol subsidies were not without purpose when they were first introduced. Most people realize that dependence on fossil fuels poses its own environmental and economic challenges that the world must somehow meet. It's essential that we explore every possible option for developing renewable, sustainable, safe, and environmentally friendly sources of energy, including solar, wind, and nuclear power, as well as potential new sources, such as geothermal energy, energy from the ocean tides, and fusion power.

When oil was priced at $20 a barrel, subsidizing ethanol to make it competitive had a certain logic. But these same subsidies cannot be justified when oil is at over $50 a barrel—nor can the continuing subsidies for oil production enjoyed by large, highly profitable firms like ExxonMobil. Both sets of subsidies distort markets; lead to unintended ecological, social, and economic consequences; and should be phased out as quickly as possible. Otherwise, they will continue to drive up the price of basic foodstuffs both directly and indirectly, including by diverting farmland and other agricultural resources to the production of fuel rather than food.

In Bangladesh, we have found that small-scale uses of non-crop biofuels can be quite effective and benign. For example, Grameen Shakti, an independent sister company in the family of Grameen companies that specializes in producing renewable energy, has helped thousands of farmers use methane gas from manure tanks as a cooking fuel and even as a source of electricity. But it is a mistake to closely link the global food and energy markets through reliance on ethanol to run the world's motor vehicles. In practice, it sacrifices the food needs of the poor to the transportation needs of the relatively well-to-do. As Robin Maynard of the U.K. Soil Association points out, the same amount of grain needed to fill the tank of a sports utility vehicle could be used to feed a person *for an entire year*.[2] This is not a fair or sustainable tradeoff.

Increased demand for meat, thanks to changing living standards, has also distorted food price structures and contributed to worldwide food shortages. Growing prosperity in some of the world's poorest nations is, of course, a wonderful thing. Over the past three decades, millions of people have been able to lift themselves out of poverty thanks to increased access to free markets, technological developments, and programs such as microcredit, which make capital for investments available to those who were once shut out of the capitalist system.

But prosperity is bringing its own challenges. The amount of meat eaten by the typical Chinese citizen has increased from twenty kilograms per year in 1958 to more than fifty kilograms today. Similar increases have been seen in other large countries such as India, Indonesia, and Bangladesh, which together with China make up nearly half the world's population. Not only can more and more people in these countries now afford meat, but they are shifting to meat (and away from more traditional, low-meat diets) as part of their adoption of a "modern," prosperous lifestyle.

Unfortunately, meat-eating is a relatively inefficient use of natural resources, as the number of nutritious calories delivered by meat is far lower than the calories humans can enjoy through direct intake of grains. Yet today, more and more grain and other foodstuffs are being used to feed cattle rather than human beings. By some measures, up to a third of the world's grain production, as well as third of the global fish catch, is being used to feed livestock. And more and more of the planet's farmlands are being diverted from the production of food for human consumption and toward the growing of grains for cattle feed, adding several costly steps to the process by which human life will ultimately be sustained.

As a result of dysfunctional agricultural choices such as the decision to shift land use toward ethanol and meat production, even basic foods are becoming more expensive. In countries where actual food shortages have not yet struck, many people are going hungry because they simply can't afford the food they see on the shelves in local shops.

There are still other factors worsening the current food crisis for the developing nations. One of these is the growing difficulty for farmers in poor nations to compete in the increasingly global food markets. As one analyst has explained,

> In theory, the growing importance of traders and supermarkets ought to make farmers more responsive to changes in prices and consumer tastes. In some places, that is the case. But supermarkets need uniform

quality, minimum large quantities and high standards of hygiene, which the average smallholder [that is, a farmer on a small plot of land] in a poor country is ill equipped to provide. So traders and supermarkets may benefit commercial farmers more than smallholders.[3]

In effect, farmers in the developing nations are suffering from the necessity to compete against large-scale producers in the developed nations. It's a one-sided battle that, so far, has led to devastating results for the poor farmers of the world.

Increasing corporate control of agricultural resources is also harming farmers in the developing world. As large agribusinesses take near-monopoly control over seed stocks as well as control over supplies of costly synthetic fertilizers and pesticides, more and more small farms are driven out of business, unable to afford the supplies they need to compete in the new global food market. The rising cost of oil is a significant factor here too. For example, many fertilizers are petroleum-based, which means that every increase in the cost of a barrel of oil drives up the cost of fertilizer. The World Bank reports that over the past five years, fertilizer prices have risen by 150 percent. Of course, high oil prices also drive up the cost of irrigating, running farm equipment, delivering goods to market, and shipping foods to and from processing plants.

As a result, farmers in the global South are becoming less and less economically viable, and the countries in which they live are becoming more dependent on food imports. And the vicious cycle of changes continues to make matters worse. As economically strapped farmers in the developing nations abandon the countryside and move to the cities in search of work, urban sprawl converts more and more land from farmland to residential or factory use. Traditional farming cultures become impoverished or lost altogether, making it more and more difficult for these nations to feed themselves.

It's difficult to convey the despair to which many once-proud farmers in countries throughout Asia, Africa, and Latin America have been driven. Shockingly, during the year 2007 alone, more than 25,000 farmers committed suicide in India, a stark comment on the depth of the problem.

All of these economic and social problems are growing worse just as global environmental trends are threatening the future of agriculture around the world. Climate change, drought, and deforestation are turning vast areas that were once fertile farmlands into deserts. The UN reports that, every year, an area equivalent to the entire country of Ukraine is lost for farming because of climate change. What's more, if current global warming trends continue, over the next century, rising sea levels can be expected to flood almost one-third of the world's farmland. It is easy to imagine what is happening to Bangladesh, the world's most densely populated country and a flat country with twenty percent of its land less than one meter above sea level. As the water level keeps rising, it is an emerging case of environmental disaster turning immediately into human disaster.

These interlocking trends are making it more and more difficult to feed the poorest of the poor worldwide. They will reduce the prospect of achieving many of the Millennium Development Goals set by the United Nations in 2000, unless both immediate short-term steps and long-term reform projects are undertaken.

Thankfully, some short-term emergency steps have been put into place. UN Secretary-General Ban Ki-moon deserves credit for convening the leaders of twenty-seven UN agencies and programs in mid-2008 to organize a coordinated response to the immediate crisis. They established a high-level task force under Ban's leadership, with sound immediate objectives. World leaders pledged some $30 billion for emergency food relief. This was an important step in the right direction.

In the intermediate term, we must ensure that farmers are equipped to produce the next harvest and are given the tools and systems they need to compete on reasonable terms in the local, regional, and international markets. As we've noted, farmers in many areas cannot afford seeds to plant or natural gas-based fertilizer, whose price has risen along with the price of oil. They need help in order to be able to feed themselves, their families, and the communities in which they live.

This will require a financial investment by the nations of the developed world. There are sources for the needed funds that seem reasonable and accessible. For example, the International Fund for Agricultural Development is already delivering $200 million to poor farmers in the countries most affected by the crisis to boost food production. The Food and Agriculture Organization needs an additional $1.7 billion to help provide seed and fertilizer. The World Bank is doubling its lending for agriculture in Africa over the next year to $800 million and is considering a new rapid financing facility for grant support to especially fragile, poor countries and quicker, more flexible financing for others.

These decisions were taken when the financial crisis was yet to devastate the world economy. But even for tackling the "silent tsunami" of hunger alone, these sums are insignificant, particularly considering the fact that the food price rise has helped some governments earn more money, or save money. In the United States alone, high food prices have been a boon to farmers and have saved the government billions of dollars in crop support payments. Rich nations that have benefited from high food prices in this way can afford to help their less fortunate sisters and brothers.

The poor of the world are now suffering under the combined effect of two giant crises. World leaders' concern for the emergency situation on the financial front is quite understandable. But while financial crisis should get the topmost priority, it should not be considered a problem of high finance only. This narrow view of the financial crisis is likely to create global social and political problems. The human aspect of the financial crisis must be integrated into all policy packages. The appropriate thing would be to treat the financial crisis and the food crisis as one crisis, because both are linked together. So far, governments have kept themselves busy in coming up with supersized bailout packages for the institutions that were responsible for creating this combined crisis, but no bailout package of any size has even been discussed for the victims of the crisis.

I have been urging that this mega-crisis be taken as a mega-opportunity to redesign the world's entire economic system, making it an inclusive system that will ensure availability of financial services to all people, even beggars. While microcredit has demonstrated how banking can be provided to the poorest women anywhere in the world without collateral and without lawyers, no system has emerged yet to provide sound, nonsubsidized agricultural financing for small holders in the developing world. In the new design of the financial system, this must get a high priority; innovation must be encouraged in this direction.

Any future bailout packages aimed at addressing the combined crises should include funds for the global expansion of microcredit to help the poor people create self-employment when jobs and incomes are lost; agricultural credit and insurance; the creation of regional and national food reserves; and support for agricultural research aimed at creating Green Revolution II as quickly as possible.

In the new design of the economic system, *social business* should also be given an important place. Social business is a new category of business that I have been advocating to fill the gaps left behind by the profit-maximizing private sector. A social business is a business created with no intention of earning personal profit and dedicated entirely to one or more specific social goals—a non-loss, non-dividend company with social objectives. The bailout package for the world's bottom three billion should include funds for investing in social businesses, perhaps through the creation of a Social Business Fund. Such a fund could extend loans and equity to social businesses addressing the following issues:

- Expansion of microcredit programs
- Agricultural credit
- Other programs to support small-scale agriculture (such as local, national, and international marketing programs; storage facilities; introduction of new technology; insurance; price and wage programs, and so on)
- Health care and health insurance

In the longer run, we need to look closely at the conditions that farmers in the developing world must struggle against in order to survive economically, and make sure that the rules of the marketplace don't discriminate unfairly against them. This isn't a matter of tilting the playing field in their favor but rather of reducing circumstances that currently make it almost impossible for the small players to survive. In the words of Olivier De Schutter, the UN's Special Rapporteur on the right to food:

Smallholders should be helped by reinforcing their ability to produce while at the same time protecting them from the consequences of volatile international prices and the risk of unfair competition from agricultural producers in industrialized countries who benefit from massive government subsidies. Other means include strengthening their ability to negotiate prices with the large agri-business firms which impose their prices on producers, and facilitating more environmentally friendly forms of agricultural production, by the use of inputs less dependent on the price of oil or on the expectations of companies holding patents on plant varieties.[4]

As part of the effort to reform the game so that small producers in the world's poorer nations can participate, crop subsidies and export controls in the rich nations that distort markets and raise prices should be eliminated.

Most important, the current mega-crisis should not distract the attention of the world leaders from the world's search for long-term global solutions to poverty and environmental protection. Instead, they should see this as an opportunity to integrate such solutions into a reformed economic system for the planet. For example, we should continue efforts to move to second-generation biofuels made from waste materials and nonfood crops without displacing land used for food production. Even the limited amount of biofuels on the market today has been credited with helping to keep the price of oil from spiraling further out of control, and next-generation fuels can be economically advantageous for poor countries with much less effect on food production. As bad as the impact of high food prices has been, the impact of high oil prices has been worse, devastating poor countries that have no indigenous source of energy supply, erasing all the benefits of international debt relief and more.

In addition, the world must develop a new system of long-term investments in agriculture. A new Green Revolution is required to meet the global demand for food, even as climate change is increasing the stresses on agriculture. More productive crops are needed, as well as new plant strains that are drought-resistant and salt-tolerant. The Consultative Group on International Agricultural Research must be strengthened to help lead these efforts, and the work of nonprofit organizations that are spearheading research and development needs to be supported. They should be encouraged and funded to collaborate with social businesses to extend their research to the farmers' benefit.

While developing countries welcome foreign direct investment, one has to be careful about foreign direct investment in agriculture. The

worry here is that these investments are ultimately bound to serve corporate interests rather than the real needs of the poorest people. Care must
be taken to ensure that the programs we invest in are sustainable and
promote the long-term food independence of the developing-world nations they are intended to help. Appropriate technologies need to be developed and applied; local people need to take the lead in planning and
implementing agricultural programs; and social and political conditions
must be addressed so that, for example, access to such basic necessities as
land and water are guaranteed for local farmers. International social
businesses may need to be created to provide technology and finance to
help agriculture and ensure that these goals are met.

The current multiple crises now troubling the world offer us all a
valuable lesson in the interconnectedness of the human family. The fate
of Lehman Brothers and that of our sisters working in the garment factories in Bangladesh are linked together. The fates of a rice farmer in
Bangladesh, a maize farmer in Mexico, and a maize farmer in Iowa are
all intertwined, and while short-term trends may appear to benefit a few
of us at the expense of many others, in the long run, only policies that
will allow *all* the peoples of the world to share their progress are truly
sustainable.

In the coming months, the multiple crises will reveal more of their
ramifications in both economic and human terms. The United States
has to play a major role in bringing the world together to face this crisis
in a well-planned and well-managed way so that similar global crises can
be avoided in the future. Under the able leadership of America's new
president, the world is hoping that the United States will lead the search
for solutions to all the outstanding and pressing issues of poverty,
hunger, health, environment, and financial meltdown.

# Another Take | THE SCOPE OF THE WORLD FOOD CRISIS

## By FoodFirst Information and Action Network

The FoodFirst Information and Action Network (FIAN) is an international human rights organization that advocates for the right to food. With national sections and individual members in more than fifty countries, FIAN is a not-for-profit organization without any religious or political affiliation and has consultative status to the United Nations. FIAN stands up against unjust and oppressive practices that prevent people from feeding themselves, including gender discrimination and other forms of exclusion. For more information, visit FIAN's website at http://www.fian.org.

This selection offer's FIAN's analysis of the seriousness of the world food crisis along with its immediate and deep-rooted causes.

Beginning in 2006, the world started to witness steady rises in the cost of food prices, which reached dramatic levels in 2007 and early 2008, creating both political and economic unrest in countries where severe poverty is widespread. As early as 2006, the world food prices for basic staples such as rice, wheat, corn, and soybeans as much as doubled. In 2008, protests broke out in more than forty countries as the result of soaring food prices. This situation is referred to as the World Food Crisis.

The World Food Crisis is just the tip of the iceberg of an ultimately structural crisis. The number of chronically malnourished people had already risen from 823 million in the mid-1990s to 843 million in 2006. The figure of those that suffer from hunger has now increased to more than 900 million people.

Hunger is mainly a rural phenomenon; eighty percent of hungry people in the world live in rural areas where food is produced, and the majority of them are smallholder peasants. Due to ill-conceived agricultural and trade policies and the neglect of smallholder agriculture, the right to food of many, especially that of peasant farmers, has been violated on a massive scale.

## CAUSES OF THE CRISIS

Initial causes of the price hikes in 2006 included unseasonable droughts in grain producing countries such as Australia and the United States, which can be seen as precursors of probable future disasters caused by climate change. Population growth, urbanization, and the increased purchasing power of middle-class populations in India and China have contributed to an increase in the consumption of meat and dairy products, and thus the use of animal feed. The threefold rise in oil prices since 2000 further heightened the costs of industrial agriculture inputs (fertilizers, food transportation).

A major cause is also certainly the search for alternative energy sources and the increasing use of agrofuels, which lead to competition with food production. Ethanol production in the United States alone accounts for more than ten percent of the worldwide corn production and was undoubtedly an important element in the soaring prices of food staples. There is also evidence that speculation played an important role in the price hikes.

Moreover, the crisis is deeply rooted in decades of misguided international and national policies, decided and implemented under the auspices of Bretton Woods Institutions such as the International Monetary Fund (IMF) and World Bank and, more recently, the World Trade Organization (WTO). These policies have undermined the policy spaces for states to respect, protect, and fulfill the human right to adequate food.

## CONTENTIOUS IMPLICATIONS AND SOLUTIONS

Far more contentious than the causes for the World Food Crisis are the implications of the price increases and the resulting political strategies and solutions. If the globally low prices of raw materials present in the 1960s were considered to be one of the most significant hindrances to development, then it would follow that high prices for commodities would be a solution. Why then do we not hear any shouts of joy coming from the increase in prices of commodities, but instead widespread protest? A closer look at states hardest hit by the crisis is illuminating. Olivier De Schutter, the UN Special Rapporteur for the Right to Food since May 2008, determines that "[t]oday's crisis is particularly disturbing for the net importers of food. Most African countries fall into this category, not least due to the liberalisation of the agrarian trade, which was imposed on them in the course of the structural adjustment measures in the 1980s and 1990s."

A good example of the effects of trade liberalization can be found in rice. Due to the opening of markets and the eradication of public services in the agricultural field, Honduras has been repeatedly affected by import surges of rice from the United States since around 1992. These economic policies had a far greater impact on domestic rice production than natural disasters such as Hurricane Mitch. While there were a reported 25,000 rice farmers in Honduras at the end of the 1980s, official statistics today estimate that there are fewer than 1,300. Between 1990 and 2000, rice production sank from 47,300 to 7,200 tons. The price increase of rice between 2007 and the middle of 2008 for the resulting necessary imports of rice has had fatal consequences. On the one hand, the cost increase in imports can give domestic farmers a chance to compete and win back market shares. On the other hand, it is hardly to be expected that domestic rice production will be available to make up for the shattered supply deficit in the short term and bring the price of rice back to an affordable level. In this respect, Honduras is no exception but is rather an example of a policy that allowed for the replacement of domestic rice production with the import of a cheap commodity.

In Haiti in 1995, under pressure from the IMF, import duties were lowered from thirty-five percent to three percent, and between 1992 and 2003 imports rose by 150 percent, which resulted in widespread protests. It is no coincidence that the protests in import-dependent countries like Haiti stem from a particular dynamic. When the world market prices for food rise, this is reflected directly in the domestic consumer prices, and the result is empty plates for the poor. As if this were not enough, according to the FAO, the total costs for imported food commodities in the poorest and lowest income nations doubled from 2000 to 2007. This is a heavy load to bear for the national budgets of these countries, and many have witnessed disastrous effects with regard to public expenditures in agriculture and social services.

## REACTIONS OF THE INTERNATIONAL COMMUNITY

Since April 2008, the reaction of the international community to the food crisis has been coordinated by the High Level Task Force on the Global Food Crisis (HLTF), which was initiated by UN Secretary-General Ban Ki-moon and is composed of all UN organizations dealing with food and agriculture issues, as well as the World Bank, the IMF, and the WTO. In July 2008, the HLTF released a Comprehensive Framework of Action (CFA), which is meant to set out the joint position of HLTF members on proposed action to overcome the food crisis.

Although the CFA suggests a review of trade and taxation policies, it forecloses the result: more deregulation at all levels, especially the reduction of tariffs, subsidies, and export restrictions. The CFA condemns export restrictions as one of the main reasons for the food crisis, without distinction or consideration of circumstances that might justify the use of such instruments in a given country in order to secure stable domestic food prices for the poor. The announcement made by the HLTF of a general lobby for trade deregulation, under the leadership of the World Bank and the IMF, raises concerns that the CFA might even lead to a further intensification of the food crisis.

Evidence of numerous studies shows that tariff reduction, among other factors, has often caused import surges of food and thereby heavily reduced local market access, incomes, and food security of smallholder farmers. In the cases of rice farmers in Ghana, Honduras, and Indonesia, as well as tomato and chicken farmers in Ghana, the right to food has clearly been violated through the reduction of import protection. While tariff reductions might be appropriate as a temporary measure to secure necessary food imports in less developed countries in times of soaring food prices, in most cases it is not an adequate strategy to establish food security and the realization of the right to food in the long run. Further trade deregulation would rather increase imports and thereby suffocate current efforts to revive domestic and smallholder-led food production. It would increase import dependency of poor countries and make them even more vulnerable to price fluctuations in the international markets.

## PROPOSALS FOR A SUSTAINABLE AGRICULTURE

The dominant view at present is that a revival of agriculture is necessary. For example, the Philippine government, after neglecting domestic rice production for decades in favor of exports such as vegetables, has made a goal to establish independence from imports by 2010. On the side of international aid givers, there is also a clear desire to revive agriculture in the global South. The crucial question is what kind of agriculture to promote. While some organizations such as the Bill and Melinda Gates Foundation call for a "New Green Revolution" for Africa, a recent study (IAASTD) supported by fifty states and approximately 400 scientists comes to a different conclusion. The emphasis of this agrarian research calls for an urgent overhaul of the agrarian system in order to fight poverty and support an ecologically conscious development system. Instead of the one-sided support of high-yield agriculture and genetic engineering to increase food production, the study recommends building on the traditional knowledge of farming communities. The most relevant recommendations of the scientists are an improved access to land, water, and sufficient seeds for marginalized small-scale farmers. The urgent need for redistributive agrarian reform becomes very clear in light of the current World Food Crisis. Those who are faring best with the rise in food prices in the developing world are those that have a sufficient amount of land at their disposal and can provide for themselves in times of crisis.

## FOR FURTHER INFORMATION

Constantin, Anne Laure. *A Time of High Prices—An Opportunity for the Rural Poor?* Minneapolis, Minn.: Institute for Agriculture and Trade Policy (IATP), 2008.

De Schutter, Olivier. "Background Note: Analysis of the World Food Crisis by the U.N. Special Rapporteur on the Right to Food." Geneva, February 5, 2008.

FAO. "Soaring Food Prices: Facts, Perspectives, Impacts, and Actions Required." Background Document for the High-Level Conference on World Food Security, Rome, June 3–5, 2008.

FAO Committee on World Food Security. "Mid-Term Review of Achieving the World Food Summit Target, Thirty-Second Session." Rome, October 30– November 4, 2006.

Human Rights Council, Seventh Session, Agenda Item 3: A/HRC/7/L.6/Rev.1. *International Assessment of Agricultural Knowledge, Science and Technology for Development (IAASTD)*, April 2008.

Issah, Mohammed. *Right to Food of Tomato and Poultry Farmers. Report of an Investigative Mission to Ghana.* Heidelberg, Germany: FIAN, Send Foundation, Both Ends, Germanwatch, and UK Food Group, 2007.

Paasch, Armin, Frank Garbers, and Thomas Hirsch, eds. *Trade Policies and Hunger: The Impact of Trade Liberalisation on the Right to Food in Rice Farming Communities in Ghana, Honduras and Indonesia.* Switzerland: Economical Advocacy Alliance, with FIAN, 2007.

von Braun, Joachim, et al. *High Food Prices: The What, Who, and the Flow of Proposed Policy Actions.* Washington, D.C.: International Food Policy Research Institute (IFPRI), 2008.

# Part III | WHAT YOU CAN DO ABOUT IT

# NINE

# WHY BOTHER?

## Michael Pollan

---

Journalist Michael Pollan is the author, most recently, of *In Defense of Food: An Eater's Manifesto*. His previous book, *The Omnivore's Dilemma: A Natural History of Four Meals* (2006), was named one of the ten best books of 2006 by *The New York Times* and *The Washington Post*. It also won the California Book Award, the Northern California Book Award, the James Beard Award for best food writing, and was a finalist for the National Book Critics Circle Award.

Pollan is also the author of *The Botany of Desire: A Plant's-Eye View of the World* (2001); *A Place of My Own* (1997); and *Second Nature* (1991). A contributing writer to the *New York Times Magazine*, Pollan is the recipient of numerous journalistic awards, including the James Beard Award for best magazine series in 2003 and the Reuters-I.U.C.N. 2000 Global Award for Environmental Journalism.

Pollan served for many years as executive editor of *Harper's Magazine* and is now the Knight Professor of Science and Environmental Journalism at the University of California at Berkeley. His articles have been anthologized in *Best American Science Writing* (2004); *Best American Essays* (1990 and 2003); and the *Norton Book of Nature Writing*. He lives in the Bay Area with his wife, the painter Judith Belzer, and their son, Isaac.

"Why Bother?" was originally published in the *The New York Times Magazine* (April 20, 2008).

---

Why bother? That really is the big question facing us as individuals hoping to do something about climate change, and it's not an easy one to answer. I don't know about you, but for me the most upsetting moment in *An Inconvenient Truth* came long after Al Gore scared the hell out of me, constructing an utterly convincing case that the very survival of life on Earth as we know it is threatened by climate change. No, the really dark moment came during the closing credits, when we are asked to . . . change our light bulbs. That's when it got really depressing. The immense disproportion between the magnitude of the problem Gore had described and the puniness of what he was asking us to do about it was enough to sink your heart.

But the drop-in-the-bucket issue is not the only problem lurking behind the "why bother" question. Let's say I do bother, big time. I turn my life upside-down, start biking to work, plant a big garden, turn down the thermostat so low I need the Jimmy Carter signature cardigan, forsake the clothes dryer for a laundry line across the yard, trade in the station wagon for a hybrid, get off the beef, go completely local. I could theoretically do all that, but what would be the point when I know full well that halfway around the world there lives my evil twin, some carbon-footprint doppelgänger in Shanghai or Chongqing who has just bought his first car (Chinese car ownership is where ours was back in 1918), is eager to swallow every bite of meat I forswear, and who's positively itching to replace every last pound of $CO_2$ I'm struggling no longer to emit. So what exactly would I have to show for all my trouble?

A sense of personal virtue, you might suggest, somewhat sheepishly. But what good is that when virtue itself is quickly becoming a term of derision? And not just on the editorial pages of *The Wall Street Journal* or on the lips of Vice President Cheney, who famously dismissed energy conservation as a "sign of personal virtue." No, even in the pages of *The New York Times* and *The New Yorker*, it seems the epithet "virtuous," when applied to an act of personal environmental responsibility, may be used only ironically. Tell me: how did it come to pass that virtue—a quality that for most of history has generally been deemed, well, a virtue—became a mark of liberal softheadedness? How peculiar, that doing the right thing by the environment—buying the hybrid, eating like a locavore—should now set you up for the Ed Begley Jr. treatment.

And even if in the face of this derision I decide I am going to bother, there arises the whole vexed question of getting it right. Is eating local or walking to work really going to reduce my carbon footprint? According to one analysis, if walking to work increases your appetite and you consume more meat or milk as a result, walking might actually emit more carbon than driving. A handful of studies have recently suggested that in certain cases under certain conditions, produce from places as far away as New Zealand might account for less carbon than comparable domestic products. True, at least one of these studies was cowritten by a representative of agribusiness interests in (surprise!) New Zealand, but even so, they make you wonder. If determining the carbon footprint of food is really this complicated, and I've got to consider not only "food miles" but also whether the food came by ship or truck and how lushly the grass grows in New Zealand, then maybe on second thought I'll just buy the imported chops at Costco, at least until the experts get their footprints sorted out.

There are so many stories we can tell ourselves to justify doing nothing, but perhaps the most insidious is that, whatever we do manage to do, it will be too little too late. Climate change is upon us, and it has arrived well ahead of schedule. Scientists' projections that seemed dire a decade ago turn out to have been unduly optimistic: the warming and the melting is occurring much faster than the models predicted. Now truly terrifying feedback loops threaten to boost the rate of change exponentially, as the shift from white ice to blue water in the Arctic absorbs more sunlight and warming soils everywhere become more biologically active, causing them to release their vast stores of carbon into the air. Have you looked into the eyes of a climate scientist recently? They look really scared.

So do you still want to talk about planting gardens?

I do.

Whatever we can do as individuals to change the way we live at this suddenly very late date does seem utterly inadequate to the challenge. It's hard to argue with Michael Specter, in a recent *New Yorker* piece on carbon footprints, when he says: "Personal choices, no matter how virtuous [N.B.!], cannot do enough. It will also take laws and money." So it will. Yet it is no less accurate or hardheaded to say that laws and money cannot do enough, either; that it will also take profound changes in the way we live. Why? Because the climate-change crisis is at its very bottom a crisis of lifestyle—of character, even. The Big Problem is nothing more nor less than the sum total of countless little everyday choices, most of them made by us (consumer spending represents seventy percent of our economy), and most of the rest of them made in the name of our needs and desires and preferences.

For us to wait for legislation or technology to solve the problem of how we're living our lives suggests we're not really serious about changing—something our politicians cannot fail to notice. They will not move until we do. Indeed, to look to leaders and experts, to laws and money and grand schemes, to save us from our predicament represents precisely the sort of thinking—passive, delegated, dependent for solutions on specialists—that helped get us into this mess in the first place. It's hard to believe that the same sort of thinking could now get us out of it.

Thirty years ago, Wendell Berry, the Kentucky farmer and writer, put forward a blunt analysis of precisely this mentality. He argued that the environmental crisis of the 1970s—an era innocent of climate change; what we would give to have back that environmental crisis!—was at its heart a crisis of character and would have to be addressed first at that level: at home, as it were. He was impatient with people who wrote checks to environmental organizations while thoughtlessly squandering fossil fuel in their everyday lives—the 1970s equivalent of people buying carbon offsets to atone for their Tahoes and Durangos. Nothing was likely to change until we healed the "split between what we think and what we do." For Berry, the "why bother" question came down to a moral imperative: "Once our personal connection to what is wrong becomes clear, then we have to choose: we can go on as before, recognizing our dishonesty and living with it the best we can, or we can begin the effort to change the way we think and live."

For Berry, the deep problem standing behind all the other problems of industrial civilization is "specialization," which he regards as the "disease of the modern character." Our society assigns us a tiny number of roles: we're producers (of one thing) at work, consumers of a great many other things the rest of the time, and then once a year or so we vote as citizens. Virtually all of our needs and desires we delegate to specialists of one kind or another—our meals to agribusiness, health to the doctor, education to the teacher, entertainment to the media, care for the environment to the environmentalist, political action to the politician.

As Adam Smith and many others have pointed out, this division of labor has given us many of the blessings of civilization. Specialization is what allows me to sit at a computer thinking about climate change. Yet this same division of labor obscures the lines of connection—and

responsibility—linking our everyday acts to their real-world consequences, making it easy for me to overlook the coal-fired power plant
that is lighting my screen, or the mountaintop in Kentucky that had to
be destroyed to provide the coal to that plant, or the streams running
crimson with heavy metals as a result.

Of course, what made this sort of specialization possible in the first
place was cheap energy. Cheap fossil fuel allows us to pay distant others to
process our food for us, to entertain us and to (try to) solve our problems,
with the result that there is very little we know how to accomplish for ourselves. Think for a moment of all the things you suddenly need to do for
yourself when the power goes out—up to and including entertaining
yourself. Think, too, about how a power failure causes your neighbors—
your community—to suddenly loom so much larger in your life. Cheap
energy allowed us to leapfrog community by making it possible to sell our
specialty over great distances as well as summon into our lives the special-
ties of countless distant others.

Here's the point: Cheap energy, which gives us climate change, fosters
precisely the mentality that makes dealing with climate change in our
own lives seem impossibly difficult. Specialists ourselves, we can no
longer imagine anyone but an expert, or anything but a new technology
or law, solving our problems. Al Gore asks us to change the light bulbs
because he probably can't imagine us doing anything much more challenging, like, say, growing some portion of our own food. We can't imagine it either, which is probably why we prefer to cross our fingers and talk
about the promise of ethanol and nuclear power—new liquids and elec-
trons to power the same old cars and houses and lives.

The "cheap-energy mind," as Wendell Berry called it, is the mind that
asks, "Why bother?" because it is helpless to imagine—much less
attempt—a different sort of life, one less divided, less reliant. Because
the cheap-energy mind translates everything into money, its proxy, it
prefers to put its faith in market-based solutions—carbon taxes and pol-
lution-trading schemes. If we could just get the incentives right, it be-
lieves, the economy will properly value everything that matters and
nudge our self-interest down the proper channels. The best we can hope
for is a greener version of the old invisible hand. Visible hands it has no
use for.

But while some such grand scheme may well be necessary, it's doubtful that it will be sufficient or that it will be politically sustainable before we've demonstrated to ourselves that change is possible. Merely to give, to spend, even to vote, is not to do, and there is so much that needs to be done—without further delay. In the judgment of James Hansen, the NASA climate scientist who began sounding the alarm on global warming twenty years ago, we have only ten years left to start cutting—not just slowing—the amount of carbon we're emitting or face a "different planet." Hansen said this more than two years ago, however; two years have gone by, and nothing of consequence has been done. So: eight years left to go and a great deal left to do.

Which brings us back to the "why bother" question and how we might better answer it. The reasons not to bother are many and compelling, at least to the cheap-energy mind. But let me offer a few admittedly tentative reasons that we might put on the other side of the scale.

If you do bother, you will set an example for other people. If enough other people bother, each one influencing yet another in a chain reaction of behavioral change, markets for all manner of green products and alternative technologies will prosper and expand. (Just look at the market for hybrid cars.) Consciousness will be raised, perhaps even changed: new moral imperatives and new taboos might take root in the culture. Driving an SUV or eating a twenty-four-ounce steak or illuminating your McMansion like an airport runway at night might come to be regarded as outrages to human conscience. Not having things might become cooler than having them. And those who did change the way they live would acquire the moral standing to demand changes in behavior from others—from other people, other corporations, even other countries.

All of this could, theoretically, happen. What I'm describing (imagining would probably be more accurate) is a process of viral social change, and change of this kind, which is nonlinear, is never something anyone can plan or predict or count on. Who knows, maybe the virus will reach all the way to Chongqing and infect my Chinese evil twin. Or not. Maybe going green will prove a passing fad and will lose steam after a few years, just as it did in the 1980s, when Ronald Reagan took down Jimmy Carter's solar panels from the roof of the White House.

Going personally green is a bet, nothing more or less, though it's one we probably all should make, even if the odds of it paying off aren't great. Sometimes you have to act as if acting will make a difference, even when you can't prove that it will. That, after all, was precisely what happened in Communist Czechoslovakia and Poland, when a handful of individuals like Vaclav Havel and Adam Michnik resolved that they would simply conduct their lives "as if" they lived in a free society. That improbable bet created a tiny space of liberty that, in time, expanded to take in, and then help take down, the whole of the Eastern bloc.

So what would be a comparable bet that the individual might make in the case of the environmental crisis? Havel himself has suggested that people begin to "conduct themselves as if they were to live on this earth forever and be answerable for its condition one day." Fair enough, but let me propose a slightly less abstract and daunting wager. The idea is to find one thing to do in your life that doesn't involve spending or voting, that may or may not virally rock the world but is real and particular (as well as symbolic) and that, come what may, will offer its own rewards. Maybe you decide to give up meat, an act that would reduce your carbon footprint by as much as a quarter. Or you could try this: determine to observe the Sabbath. For one day a week, abstain completely from economic activity: no shopping, no driving, no electronics.

But the act I want to talk about is growing some—even just a little—of your own food. Rip out your lawn, if you have one, and if you don't—if you live in a high-rise, or have a yard shrouded in shade—look into getting a plot in a community garden. Measured against the Problem We Face, planting a garden sounds pretty benign, I know, but in fact it's one of the most powerful things an individual can do—to reduce your carbon footprint, sure, but more important, to reduce your sense of dependence and dividedness: to change the cheap-energy mind.

A great many things happen when you plant a vegetable garden, some of them directly related to climate change, others indirect but related nevertheless. Growing food, we forget, comprises the original solar technology: calories produced by means of photosynthesis. Years ago the cheap-energy mind discovered that more food could be produced with less effort by replacing sunlight with fossil-fuel fertilizers and pesticides, with a result that the typical calorie of food energy in your diet now requires about ten calories of fossil-fuel energy to produce. It's estimated that the way we feed ourselves (or rather, allow ourselves to be fed) accounts for about a fifth of the greenhouse gas for which each of us is responsible.

Yet the sun still shines down on your yard, and photosynthesis still works so abundantly that in a thoughtfully organized vegetable garden (one planted from seed, nourished by compost from the kitchen, and involving not too many drives to the garden center), you can grow the proverbial free lunch—$CO_2$-free and dollar-free. This is the most-local food you can possibly eat (not to mention the freshest, tastiest, and most nutritious), with a carbon footprint so faint that even the New Zealand lamb council dares not challenge it. And while we're counting carbon, consider too your compost pile, which shrinks the heap of garbage your household needs trucked away even as it feeds your vegetables and sequesters carbon in your soil. What else? Well, you will probably notice that you're getting a pretty good workout there in your garden, burning calories without having to get into the car to drive to the gym. (It is one of the absurdities of the modern division of labor that, having replaced physical labor with fossil fuel, we now have to burn even more fossil fuel to keep our unemployed bodies in shape.) Also, by engaging both body and mind, time spent in the garden is time (and energy) subtracted from electronic forms of entertainment.

You begin to see that growing even a little of your own food is, as Wendell Berry pointed out thirty years ago, one of those solutions that, instead of begetting a new set of problems—the way "solutions" like ethanol or nuclear power inevitably do—actually beget other solutions, and not only of the kind that save carbon. Still more valuable are the habits of mind that growing a little of your own food can yield. You quickly learn that you need not be dependent on specialists to provide for yourself—that your body is still good for something and may actually be enlisted in its own support. If the experts are right, if both oil and time are running out, these are skills and habits of mind we're all very soon going to need. We may also need the food. Could gardens provide it? Well, during World War II, victory gardens supplied as much as forty percent of the produce Americans ate.

But there are sweeter reasons to plant that garden, to bother. At least in this one corner of your yard and life, you will have begun to heal the split between what you think and what you do, to commingle your identities as consumer and producer and citizen. Chances are, your garden will reengage you with your neighbors, for you will have produce to give away and the need to borrow their tools. You will have reduced the power of the cheap-energy mind by personally overcoming its most debilitating weakness: its helplessness and the fact that it can't do much of anything that doesn't involve division or subtraction. The garden's season-long transit from seed to ripe fruit—will you get a load of that zucchini?!—suggests that the operations of addition and multiplication still obtain, that the abundance of nature is not exhausted. The single greatest lesson the garden teaches is that our relationship to the planet need not be zero-sum, and that as long as the sun still shines and people still can plan and plant, think and do, we can, if we bother to try, find ways to provide for ourselves without diminishing the world.

# Another TEN STEPS TO STARTING
# Take | A COMMUNITY GARDEN

*By the American Community Gardening Association*

---

The mission of the American Community Gardening Association (ACGA) is to build community by increasing and enhancing community gardening and greening across the United States and Canada. A binational nonprofit membership organization of professionals, volunteers, and supporters of community greening in urban and rural communities, ACGA recognizes that community gardening improves people's quality of life by providing a catalyst for neighborhood and community development; stimulating social interaction; encouraging self-reliance; beautifying neighborhoods; producing nutritious food; reducing family food budgets; conserving resources; and creating opportunities for recreation, exercise, therapy, and education.

ACGA and its member organizations support community gardening by facilitating the formation and expansion of state and regional community gardening networks; developing resources in support of community gardening; and encouraging research and conducting educational programs. For more information, visit ACGA's website at http://www.communitygarden.org.

---

The following steps are adapted from the American Community Garden Association's guidelines for launching a successful community garden in your neighborhood.

1.  **Organize a meeting of interested people**

    Determine whether a garden is really needed and wanted, what kind it should be (vegetable, flower, both, organic?), whom it will involve, and who benefits. Invite neighbors, tenants, community organizations, gardening and horticultural societies, building superintendents (if it is at an apartment building)—in other words, anyone who is likely to be interested.

2. **Form a planning committee**

This group can be comprised of people who feel committed to the creation of the garden and have the time to devote to it, at least at this initial stage. Choose well-organized persons as garden coordinators. Form committees to tackle specific tasks: funding and partnerships, youth activities, construction, and communication.

3. **Identify all your resources**

Do a community asset assessment. What skills and resources already exist in the community that can aid in the garden's creation? Contact local municipal planners about possible sites, as well as horticultural societies and other local sources of information and assistance. Look within your community for people with experience in landscaping and gardening. For example, in Toronto, contact the Toronto Community Garden Network.

4. **Approach a sponsor**

Some gardens "self-support" through membership dues, but for many, a sponsor is essential for donations of tools, seeds, or money. Churches, schools, private businesses, or parks and recreation departments are all possible supporters. One garden raised money by selling "square inches" at $5 each to hundreds of sponsors.

5. **Choose a site**

Consider the amount of daily sunshine (vegetables need at least six hours a day), availability of water, and soil testing for possible pollutants. Find out who owns the land. Can the gardeners get a lease agreement for at least three years? Will public liability insurance be necessary?

6. **Prepare and develop the site**

In most cases, the land will need considerable preparation for planting. Organize volunteer work crews to clean it, gather materials, and decide on the design and plot arrangement.

7. **Organize the garden**

Members must decide how many plots are available and how they will be assigned. Allow space for storing tools, making compost, and don't forget the pathways between plots! Plant flowers or shrubs around the garden's edges to promote good will with non-gardening neighbors, passersby, and municipal authorities.

8. **Plan for children**

Consider creating a special garden just for kids—including them is essential. Children are not as interested in the size of the harvest but rather in the process of gardening. A separate area set aside for them allows them to explore the garden at their own speed.

9. **Determine rules and put them in writing**

The gardeners themselves devise the best ground rules. We are more willing to comply with rules that we have had a hand in creating. Ground rules help gardeners to know what is expected of them. Think of it as a code of behavior. Some examples of issues that are best dealt with by agreed-upon rules are the following: What are the dues? How will the money be used? How are plots assigned? Will gardeners share tools, meet regularly, handle basic maintenance?

10. **Help members keep in touch with each other**

Good communication ensures a strong community garden with active participation by all. Some ways to do this are to form a telephone tree, to create an email list, to install a rainproof bulletin board in the garden, and to have regular celebrations. Community gardens are all about creating and strengthening communities.

# TEN

# DECLARE YOUR INDEPENDENCE

## *By Joel Salatin*

Joel Salatin is a full-time farmer in Virginia's Shenandoah Valley. A third-generation alternative farmer, he returned to his family's Polyface Farm ("The Farm of Many Faces")
full-time in 1982 and continued refining and adding to his parents' ideas.

Today, Polyface Farm services more than 1,500 families, ten retail outlets, and thirty
restaurants through on-farm sales and metropolitan buying clubs with salad bar beef,
pastured poultry, eggmobile eggs, pigaerator pork, forage-based rabbits, pastured
turkey, and forestry products using relationship marketing. It has been featured in
*Smithsonian*; *National Geographic*; *Gourmet*; and countless other radio, television, and
print media and was featured in *The New York Times*–best seller *The Omnivore's
Dilemma* by food writer/guru Michael Pollan.

Salatin holds a bachelor's degree in English and writes extensively in magazines
such as *Stockman Grass Farmer*, *Acres USA*, and *American Agriculturalist*. He has also
authored six books, including four how-to books on farming as well as *Holy Cows and
Hog Heaven: The Food Buyer's Guide to Farm Friendly Food* and *Everything I Want to
Do Is Illegal: War Stories from the Local Food Front*.

Perhaps the most empowering concept in any paradigm-challenging
movement is simply opting out. The opt-out strategy can humble the
mightiest forces because it declares to one and all, "You do not control me."

The time has come for people who are ready to challenge the paradigm of factory-produced food and to return to a more natural, wholesome, and sustainable way of eating (and living) to make that declaration
to the powers that be, in business and government, that established the
existing system and continue to prop it up. It's time to opt out and simply
start eating better—right here, right now.

Impractical? Idealistic? Utopian? Not really. As I'll explain, it's actually the most realistic and effective approach to transforming a system that is slowly but surely killing us.

## WHAT HAPPENED TO FOOD?

First, why am I taking a position that many well-intentioned people might consider alarmist or extreme? Let me explain.

At the risk of stating the obvious, the unprecedented variety of bar-coded packages in today's supermarket really does not mean that our generation enjoys better food options than our predecessors. These packages, by and large, having passed through the food inspection fraternity, the industrial food fraternity, and the lethargic cheap-food-purchasing consumer fraternity, represent an incredibly narrow choice. If you took away everything with an ingredient foreign to our three trillion intestinal microflora, the shelves would be bare indeed. (I'm talking here about the incredible variety of microorganisms that live in our digestive tracts and perform an array of useful functions, including training our immune systems and producing vitamins like biotin and vitamin K.) In fact, if you just eliminated every product that would have been unavailable in 1900, almost everything would be gone, including staples that had been chemically fertilized, sprayed with pesticides, or ripened with gas.

Rather than representing newfound abundance, these packages wending their way to store shelves after spending a month in the belly of Chinese merchant marines are actually the meager offerings of a tyrannical food system. Strong words? Try buying real milk—as in raw. See if you can find meat processed in the clean open air under sterilizing sunshine. Look for pot pies made with local produce and meat. How about good old unpasteurized apple cider? Fresh cheese? Unpasteurized almonds? All these staples that our great-grandparents relished and grew healthy on have been banished from today's supermarkets.

They've been replaced by an array of pseudo-foods that did not exist a mere century ago. The food additives, preservatives, colorings, emulsifiers, corn syrups, and unpronounceable ingredients listed on the colorful packages bespeak a centralized control mindset that actually reduces the options available to fill Americans' dinner plates. Whether by intentional design or benign ignorance, the result has been the same—the criminalization and/or demonization of heritage foods.

The mindset behind this radical transformation of American eating habits expresses itself in at least a couple of ways.

One is the completely absurd argument that without industrial food, the world would starve. "How can you feed the world?" is the most common question people ask me when they tour Polyface Farm. Actually, when you consider the fact that millions of people, including many vast cities, were fed and sustained using traditional farming methods until just a few decades ago, the answer is obvious. America has traded seventy-five million buffalo, which required no tillage, petroleum, or chemicals, for a mere forty-two million head of cattle. Even with all the current chemical inputs, our production is a shadow of what it was 500 years ago. Clearly, if we returned to herbivorous principles five centuries old, we could double our meat supply. The potential for similar increases exists for other food items.

The second argument is about food safety. "How can we be sure that food produced on local farms without centralized inspection and processing is really safe to eat?" Here, too, the facts are opposite to what many people assume. The notion that indigenous food is unsafe simply has no scientific backing. Milk-borne pathogens, for example, became a significant health problem only during a narrow time period between 1900 and 1930, before refrigeration but after unprecedented urban expansion. Breweries needed to be located near metropolitan centers, and adjacent dairies fed herbivore-unfriendly brewery waste to cows. The combination created real problems that do not exist in grass-based dairies practicing good sanitation under refrigeration conditions.

Lest you think the pressure to maintain the industrialized food system is all really about food safety, consider that all the natural-food items I listed above can be given away, and the donors are considered pillars of community benevolence. But as soon as money changes hands, all these wonderful choices become "hazardous substances," guaranteed to send our neighbors to the hospital with food poisoning. Maybe it's not human health but corporate profits that are really being protected.

Furthermore, realize that many of the same power brokers (politicians and the like) encourage citizens to go out into the woods on a 70-degree fall day; gut-shoot a deer with possible variant Creutzfeld-Jacob's disease (like mad cow for deer); drag the carcass a mile through squirrel dung, sticks, and rocks; then drive parade-like through town in the blazing afternoon sun with the carcass prominently displayed on the hood of the Blazer. The hunter takes the carcass home, strings it up in the backyard tree under roosting birds for a week, then skins it out and feeds the meat to his children. This is all considered noble and wonderful, even patriotic. Safety? It's not an issue.

The question is, who decides what food is safe? In our society, the decisions are made by the same type of people who decided in the Dred Scott ruling that slaves were not human beings. Just because well-educated, credentialed experts say something does not make it true. History abounds with expert opinion that turned out to be dead wrong. Ultimately, food safety is a personal matter of choice, of conscience. In fact, if high-fructose corn syrup is hazardous to health—and certainly we could argue that it is—then half of the government-sanctioned food in supermarkets is unsafe. Mainline soft drinks would carry a warning label. Clearly, safety is a subjective matter.

## RECLAIMING FOOD FREEDOM

Once we realize that safety is a matter of personal choice, individual freedom suddenly—and appropriately—takes center stage. What could be a more basic freedom than the freedom to choose what to feed my three-trillion-member internal community?

In America I have the freedom to own guns, speak, and assemble. But what good are those freedoms if I can't choose to eat what my body wants in order to have the energy to shoot, preach, and worship? The only reason the framers of the American Constitution and Bill of Rights did not guarantee freedom of food choice was that they couldn't envision a day when neighbor-to-neighbor food commerce would be criminalized . . . when the bureaucratic-industrial food fraternity would subsidize corn syrup and create a nation of diabetes sufferers, but deny my neighbor a pound of sausage from my Thanksgiving hog killin'.

People tend to have short memories. We all assume that whatever is must be normal. Industrial food is not normal. Nothing about it is normal.

In the continuum of human history, what western civilization has done to its food in the last century represents a mere blip. It is a grand experiment on an ever-widening global scale. We have not been here before. The three trillion members of our intestinal community have not been here before. If we ate like humans have eaten for as long as anyone has kept historical records, almost nothing in the supermarket would be on the table.

A reasonable person, looking at the lack of choice we now suffer, would ask for a Food Emancipation Proclamation. Food has been enslaved by so-called inspectors that deem the most local, indigenous, heritage-based, and traditional foods unsafe and make them illegal. It has been enslaved by a host-consuming agricultural parasite called "government farm subsidies." It has been enslaved by corporate-subsidized research that declared for four decades that feeding dead cows to cows was sound science—until mad cows came to dinner.

The same criminalization is occurring on the production side. The province of Quebec has virtually outlawed outdoor poultry. Ponds, which stabilize hydrologic cycles and have forever been considered natural assets, are now considered liabilities because they encourage wild birds, which could bring avian influenza. And with the specter of a National Animal Identification System being rammed down farmers' throats, small flocks and herds are being economized right out of existence.

On our Polyface Farm nestled in Virginia's Shenandoah Valley, we have consciously opted out of the industrial production and marketing paradigms. Meat chickens move every day in floorless, portable shelters across the pasture, enjoying bugs, forage, and local grain (grown free of genetically modified organisms). Tyson-style, inhumane, fecal factory chicken houses have no place here.

The magical land-healing process we use, with cattle using mob-stocking, herbivorous, solar conversion, lignified carbon sequestration fertilization, runs opposite the grain-based feedlot system practiced by mainline industrial cattle production. We move the cows every day from paddock to paddock, allowing the forage to regenerate completely through its growth curve, metabolizing solar energy into biomass.

Our pigs aerate anaerobic, fermented bedding in the hay feeding shed, where manure, carbon, and corn create a pig delight. We actually believe that honoring and respecting the "pigness" of the pig is the first step in an ethical, moral cultural code. By contrast, today's industrial food system views pigs as merely inanimate piles of protoplasmic molecular structure to be manipulated with whatever cleverness the egocentric human mind can conceive. A society that views its plants and animals from that manipulative, egocentric, mechanistic mindset will soon come to view its citizens in the same way. How we respect and honor the least of these is how we respect and honor the greatest of these.

The industrial pig growers are even trying to find the stress gene so it can be taken out of the hog's DNA. That way the pigs can be abused but won't be stressed about it. Then they can be crammed in ever tighter quarters without cannibalizing and getting sick. In the name of all that's decent, what kind of ethics encourages such notions?

In just the last couple of decades, Americans have learned a new lexicon of squiggly Latin words: camphylobacter, lysteria, E. coli, salmonella, bovine spongiform encephalopathy, avian influenza. Whence these strange words? Nature is speaking a protest, screaming to our generation: "Enough!" The assault on biological dignity has pushed nature to the limit. Begging for mercy, its pleas go largely unheeded on Wall Street, where Conquistadors subjugating weaker species think they can forever tyrannize without an eventual payback. But the rapist will pay—eventually. You and I must bring a nurturing mentality to the table to balance the industrial food mindset.

Here at Polyface, eggmobiles follow the cows through the grazing cycle. These portable laying hen trailers allow the birds to scratch through the cows' dung and harvest newly uncovered crickets and grasshoppers, acting like a biological pasture sanitizer. This biomimicry stands in stark contrast to chickens housed beak by wattle in egg factories, never allowed to see sunshine or chase a grasshopper.

We have done all of this without money or encouragement from those who hold the reins of food power, government or private. We haven't asked for grants. We haven't asked for permission. In fact, to the shock and amazement of our urban friends, our farm is considered a Typhoid Mary by our industrial farm neighbors. Why? Because we don't medicate, vaccinate, genetically adulterate, irradiate, or exudate like they do. They fear our methods because they've been conditioned by the powers that be to fear our methods.

The point of all this is that if anyone waits for credentialed industrial experts, whether government or nongovernment, to create ecologically, nutritionally, and emotionally friendly food, they might as well get ready for a long, long wait. For example, just imagine what a grass-finished herbivore paradigm would do to the financial and power structure of America. Today, roughly seventy percent of all grains go through herbivores, which aren't supposed to eat them and, in nature, never do. If the land devoted to that production were converted to perennial prairie polycultures under indigenous biomimicry management, it would topple the grain cartel and reduce petroleum usage, chemical usage, machinery manufacture, and bovine pharmaceuticals.

Think about it. That's a lot of economic inertia resisting change. Now do you see why the Farm Bill that controls government input into our agricultural system never changes by more than about two percent every few years? Even so-called conservation measures usually end up serving the power brokers when all is said and done.

## OPTING OUT

If things are going to change, it is up to you and me to make the change. But what is the most efficacious way to make the change? Is it through legislation? Is it by picketing the World Trade Organization talks? Is it by dumping cow manure on the parking lot at McDonald's? Is it by demanding regulatory restraint over the aesthetically and aromatically repulsive industrial food system?

At the risk of being labeled simplistic, I suggest that the most efficacious way to change things is simply to declare our independence from the figurative kings in the industrial system. To make the point clear, here are the hallmarks of the industrial food system:

- Centralized production
- Mono-speciation
- Genetic manipulation
- Centralized processing
- Confined animal feeding operations
- Things that end in "cide" (Latin for death)

- Ready-to-Eat food
- Long-distance transportation
- Externalized costs—economy, society, ecology
- Pharmaceuticals
- Opaqueness
- Unpronounceable ingredients
- Supermarkets
- Fancy packaging
- High fructose corn syrup
- High liability insurance
- "No Trespassing" signs

Reviewing this list shows the magnitude and far-reaching power of the industrial food system. I contend that it will not move. Entrenched paradigms never move . . . until outside forces move them. And those forces always come from the bottom up. The people who sit on the throne tend to like things the way they are. They have no reason to change until they are forced to do so.

The most powerful force you and I can exert on the system is to opt out. Just declare that we will not participate. Resistance movements from the antislavery movement to women's suffrage to sustainable agriculture always have and always will begin with opt-out resistance to the status quo. And seldom does an issue present itself with such a daily—in fact, thrice daily—opportunity to opt out.

Perhaps the best analogy in recent history is the home-school movement. In the late 1970s, as more families began opting out of institutional educational settings, credentialed educational experts warned us about the jails and mental asylums we'd have to build to handle the educationally and socially deprived children that home-schooling would produce. Many parents went to jail for violating school truancy laws. A quarter-century later, of course, the paranoid predictions are universally recognized as wrong. Not everyone opts for home-schooling, but the option must be available for those who want it. In the same way, an opt-out food movement will eventually show the Henny Penny food police just how wrong they are.

## LEARN TO COOK AGAIN

I think the opt-out strategy involves at least four basic ideas.

First, we must rediscover our kitchens. Never has a culture spent more to remodel and techno-glitz its kitchens, but at the same time been more lost as to where the kitchen is and what it's for. As a culture, we don't cook any more. Americans consume nearly a quarter of all their food in their cars, for crying out loud. Americans graze through the kitchen, popping precooked, heat-and-eat, bar-coded packages into the microwave for eating-on-the-run.

That treatment doesn't work with real food. Real heritage food needs to be sliced, peeled, sautéed, marinated, puréed, and a host of other things that require true culinary skills. Back in the early 1980s when our farm began selling pastured poultry, nobody even asked for boneless, skinless breast. To be perfectly sexist, every mom knew how to cut up a chicken. That was generic cultural mom information. Today, half of the moms don't know that a chicken even has bones.

I was delivering to one of our buying club drops a couple of months ago, and one of the ladies discreetly pulled me aside and asked: "How do you make a hamburger?" I thought I'd misunderstood, and asked her to repeat the question. I bent my ear down close to hear her sheepishly repeat the same question. I looked at her incredulously and asked: "Are you kidding?"

"My husband and I have been vegetarians. But now that we realize we can save the world by eating grass-based livestock, we're eating meat, and he wants a hamburger. But I don't know how to make it." This was an upper-middle-income, college-educated, bright, intelligent woman.

The indigenous knowledge base surrounding food is largely gone. When "scratch" cooking means actually opening a can, and when church and family reunion potlucks include buckets of Kentucky Fried Chicken, you know our culture has suffered a culinary information implosion. Big time. Indeed, according to marketing surveys, roughly seventy percent of Americans have no idea what they are having for supper at 4:00 p.m. That's scary.

Whatever happened to planning the week's menus? We still do that at our house. In the summer, our Polyface interns and apprentices enjoy creating a potluck for all of us Salatins every Saturday evening. All week they connive to plan the meal. It develops throughout the week, morphs into what is available locally and seasonally, and always culminates in a fellowship feast.

As a culture, if all we did was rediscover our kitchens and quit buying prepared foods, it would fundamentally change the industrial food system. The reason I'm leading this discussion with that option is because too often the foodies and greenies seem to put the onus for change on the backs of farmers. But this is a team effort, and since farmers do not even merit Census Bureau recognition, non-farmers must ante up to the responsibility for the change. And both moms and dads need to reclaim the basic food preparation knowledge that was once the natural inheritance of every human being.

## BUY LOCAL

After rediscovering your kitchen, the next opt-out strategy is to purchase as directly as possible from your local farmer. If the money pouring into industrial food dried up tomorrow, that system would cease to exist. Sounds easy, doesn't it? Actually, it is. It doesn't take any legislation, regulation, taxes, agencies, or programs. As the money flows to local producers, more producers will join them. The only reason the local food system is still minuscule is because few people patronize it.

Even organics have been largely co-opted by industrial systems. Go to a food co-op drop, and you'll find that more than half the dollars are being spent for organic corn chips, treats, and snacks. From far away.

Just for fun, close your eyes and imagine walking down the aisle of your nearby Wal-Mart or Whole Foods. Make a note of each item as you walk by and think about what could be grown within one hundred miles of that venue. I recommend this exercise when speaking at conferences all over the world, and it's astounding the effect it has on people. As humans, we tend to get mired in the sheer monstrosity of it all. But if we break it down into little bits, suddenly the job seems doable. Can milk be produced within one hundred miles of you? Eggs? Tomatoes? Why not?

Not everything can be grown locally, but the lion's share of what you eat certainly can. I was recently in the San Joaquin Valley looking at almonds—square miles of almonds. Some eighty-five percent of all the world's almonds are grown in that area. Why not grow a variety of things for the people of Los Angeles instead? My goodness, if you're going to irrigate anyway, why not grow things that will be eaten locally rather than things that will be shipped to some far corner of the world? Why indeed? Because most people aren't asking for local. Los Angeles is buying peas from China so almonds can be shipped to China.

Plenty of venues exist for close exchange to happen. Farmers' markets are a big and growing part of this movement. They provide a social atmosphere and a wide variety of fare. Too often, however, their politics and regulations stifle vendors. And they aren't open every day for the convenience of shoppers.

Community-supported agriculture (CSA) is a shared-risk investment that answers some of the tax and liability issues surrounding food commerce. Patrons invest in a portion of the farm's products and receive a share every week during the season. The drawback is the paperwork and lack of patron choice.

Food boutiques or niche retail facades are gradually filling a necessary role because most farmers' markets are not open daily. The price markup may be more, but the convenience is real. These allow farmers to drop off products quickly and go back to farming or other errands. Probably the biggest challenge with these venues is their overhead relative to scale.

Farmgate sales, especially near cities, are wonderful retail opportunities. Obviously, traveling to the farm has its drawbacks, but actually visiting the farm creates an accountability and transparency that are hard to achieve in any other venue. To acquire food on the farmer's own turf creates a connection, relationship, and memory that heighten the intimate dining experience. The biggest hurdle is zoning laws that often do not allow neighbors to collaboratively sell. (My book *Everything I Want to Do Is Illegal* details the local food hurdles in greater detail.)

Metropolitan buying clubs (MBCs) are developing rapidly as a new local marketing and distribution venue. Using the Internet as a farmer-to-patron real-time communication avenue, this scheme offers scheduled drops in urban areas. Patrons order via the Internet from an inventory supplied by one or more farms. Drop points in their neighborhoods offer easy access. Farmers do not have farmers' market politics or regulations to deal with, or sales commissions to pay. This transaction is highly efficient because it is nonspeculative—everything that goes on the delivery vehicle is preordered, and nothing comes back to the farm. Customizing each delivery's inventory for seasonal availability offers flexibility and an info-dense menu.

Many people ask, "Where do I find local food, or a farmer?" My answer: "They are all around. If you will put as much time into sourcing your local food as many people put into picketing and political posturing, you will discover a whole world that Wall Street doesn't know exists." I am a firm believer in the Chinese proverb: "When the student is ready, the teacher will appear." This nonindustrial food system lurks below the radar in every locality. If you seek, you will find.

## BUY WHAT'S IN SEASON

After discovering your kitchen and finding your farmer, the third opt-out procedure is to eat seasonally. This includes "laying by" for the off season. Eating seasonally does not mean denying yourself tomatoes in January if you live in New Hampshire. It means procuring the mountains of late-season tomatoes thrown away each year and canning, freezing, or dehydrating them for winter use.

In our basement, hundreds of quarts of canned summer produce line the pantry shelves. Green beans, yellow squash, applesauce, pickled beets, pickles, relish, and a host of other delicacies await off-season menus. I realize this takes time, but it's the way for all of us to share bioregional rhythms. To refuse to join this natural food ebb and flow is to deny connectedness. And this indifference to life around us creates a jaundiced view of our ecological nest and our responsibilities within it.

For the first time in human history, a person can move into a community, build a house out of outsourced material, heat it with outsourced energy, hook up to water from an unknown source, send waste out a pipe somewhere else, and eat food from an unknown source. In other words, in modern America we can live without any regard to the ecological life raft that undergirds us. Perhaps that is why many of us have become indifferent to nature's cry.

The most unnatural characteristic of the industrial food system is the notion that the same food items should be available everywhere at once at all times. To have empty grocery shelves during inventory downtime is unthinkable in the supermarket world. When we refuse to participate in the nonseasonal game, it strikes a heavy blow to the infrastructure, pipeline, distribution system, and ecological assault that upholds industrial food.

## PLANT A GARDEN

My final recommendation for declaring your food independence is to grow some of your own. I am constantly amazed at the creativity shown by urban-dwellers who physically embody their opt-out decision by growing something themselves. For some, it may be a community garden where neighbors work together to grow tomatoes, beans, and squash. For others, it may be three or four laying hens in an apartment. Shocking? Why? As a culture, we think nothing of having exotic tropical birds in city apartments. Why not use that space for something productive, like egg layers? Feed them kitchen scraps and gather fresh eggs every day.

Did someone mention something about ordinances? Forget them. Do it anyway. Defy. Don't comply. People who think nothing of driving around Washington, D.C., at eighty miles an hour in a fifty-five speed limit zone often go apoplectic at the thought of defying a zoning- or building-code ordinance. The secret reality is that the government is out of money and can't hire enough bureaucrats to check up on everybody anyway. So we all need to just begin opting out and it will be like five lanes of speeders on the beltway—who do you stop?

Have you ever wanted to have a cottage business producing that wonderful soup, pot pie, or baked item your grandmother used to make? Well, go ahead and make it, sell it to your neighbors and friends at church or garden club. Food safety laws? Forget them. People getting sick from food aren't getting it from their neighbors; they are getting it from USDA-approved, industrially produced, irradiated, amalgamated, adulterated, reconstituted, extruded, pseudo-food laced with preservatives, dyes, and high fructose corn syrup.

If you live in a condominium complex, approach the landlord about taking over a patch for a garden. Plant edible landscaping. If all the campuses in Silicon Valley would plant edible varieties instead of high-maintenance ornamentals, their irrigation water would actually be put to ecological use instead of just feeding hedge clippers and lawn mower engines. Urban garden projects are taking over abandoned lots, and that is a good thing. We need to see more of that. Schools can produce their own food. Instead of hiring Chemlawn, how about running pastured poultry across the yard? Students can butcher the chickens and learn about the death-life-death-life cycle.

Clearly, so much can be done right here, right now, with what you and I have. The question is not, "What can I force someone else to do?" The question is, "What am I doing today to opt out of the industrial food system?" For some, it may be having one family sit-down, locally-sourced meal a week. That's fine. We haven't gotten where we've gotten overnight, and we certainly won't extract ourselves from where we are overnight.

But we must stop feeling like victims and adopt a proactive stance. The power of many individual right actions will then compound to create a different culture. Our children deserve it. And the earthworms will love us—along with the rest of the planet.

# Another Take | QUESTIONS FOR A FARMER

## By Sustainable Table

---

Sustainable Table is a website that was created in 2003 by the nonprofit organization GrassRoots Action Center for the Environment (GRACE) to help consumers understand the problems with our food supply and offer viable solutions and alternatives. Rather than be overwhelmed by the problems created by our industrial agricultural system, Sustainable Table celebrates the joy of food and eating.

Sustainable Table is home to the Eat Well Guide, an online directory of sustainably raised meat, poultry, dairy, and eggs from farms, stores, restaurants, bed & breakfasts, and other outlets in the United States and Canada. Consumers enter their zip or postal code to find wholesome products available locally or when traveling. The Eat Well Guide currently hosts nearly 9,000 entries, with new outlets added daily, as well as a list of local and national organizations working on sustainable food issues.

Visitors to Sustainable Table can also access the critically acclaimed, award-winning Meatrix movies—*The Meatrix*, *The Meatrix II: Revolting*, and *The Meatrix II½*.

In "Questions for a Farmer," the experts at Sustainable Table offer questions consumers can use to determine whether or not the organic foods they're considering buying have really been produced in a sustainable fashion. To learn more, visit the Sustainable Table website at http://www.sustainabletable.org.

---

Sometimes, the hardest part of learning something new is knowing what questions to ask. Below we've provided you with some questions we might ask a farmer about their sustainable food-production practices, along with the answers that you should be listening for. Each group of questions below is designed for use with a particular type of farmer. We've repeated questions as appropriate from one heading to the next, so it will be easy for you to copy that section and pop it in your purse or pocket when you visit with a farmer or retailer.

Sustainable farmers are very open about how they raise their animals. If you're not on their farm asking these questions, ask to visit and see exactly how the animals are raised. The vast majority of farmers would love to have you stop by for a visit! Alternatively, if your meat came from a company that distributes products raised by family ranchers, ask for their written protocol (the standards for exactly how the animal was raised). If they won't, you might want to think twice about buying their products.

## QUESTIONS FOR A FARMER: BEEF

1. **Was the animal raised on pasture? How was the animal raised?**

   When animals are raised on pasture or in fields, they graze outdoors. Cows belong to a group of animals called ruminants, and their stomachs are designed to digest grasses, so it's best to find farmers who let their cows out to pasture.

2. **Was the animal fed anything else besides grass?**

   Find out if the farmer fed the animal any supplements, by-products, or additional type of feed. On factory farms, the animals might be fed such things as poultry manure and feathers, cement dust, rotten and outdated food, and other unsavory products.

   Animals are also fed animal by-products, even since the 2003 mad cow disease discovery in the United States. There are efforts to stop the loopholes where cows can be fed back to cows (as of January 2009, the loopholes still existed). But even if all the loopholes are closed, other animals can still be ground up and fed to cows (who, remember, are vegetarians by nature). Supplements often contain animal fat and protein.

   The only way to be sure there are no animal products in the feed being given to the animal is to know that the animal had a one hundred percent vegetarian diet or was given one hundred percent vegetarian feed, though a high amount of corn in the animal's diet can make it sick.

3. **How was the animal finished?**

"Finishing" is the process an animal goes through as it's being readied for slaughter. If an animal is finished on pasture, it means it ate grasses up until slaughter.

If an animal is finished on grain (corn is the hardest to digest), it means that for a certain amount of time before processing, it was fed grain. The grain gives the meat the marbling texture that most consumers are used to, but providing only grain to a cow can make it sick.

If the cow was finished on grain, you can also ask: (a) How long was the animal finished on grain? (b) Was it finished in a feedlot? If so, for how long?

The issue of finishing is complex because there are different beliefs among sustainable farmers. Some believe that animals should only be fed grasses because they can't digest grains properly. On factory farm feedlots—where animals eat only grains, animal by-products, and other unsavory substances—they often get sick because their stomachs can't properly digest the food.

Many sustainable farmers finish their animals on grain, but they employ what we call "grain supplemented." The animals are fed a mixture of grain and grasses—they are not forced to eat grain, but it is provided for them in small amounts in the field. The animals can eat as much grain as they want, and will stop when they no longer wish to eat any. They are also given a controlled amount so they cannot over-eat and make themselves sick. The farmer in no way forces the animals to eat grain, but they are fed it and do eat it. The farmers who employ this method say that their animals do not get sick.

Another question to consider is how old the animals were when they started eating grain. Many commercial feedlots feed corn to newly weaned calves whose stomachs are not mature enough to digest the grain. More sustainable finishers will wait until an animal is fourteen to eighteen months old before starting it on grain. These finishers will also seek to provide a wider mix of grains, rather than give only the cheapest available (usually corn, which is harder to digest but tends to be very low-cost).

The issue of a feedlot is also tricky—feedlots are where most animals go to for a short time before they are slaughtered. It's basically the holding pen before the animal goes to the processing plant. Some go for several days—others are there for six months. We believe that animals should not live in a feedlot at all, but some farmers have no choice except to send their animals to a feedlot for a certain amount of time because there are times of the year when grass is simply not available— even many "grass-fed" beef cattle have spent a considerable portion of their lives in a feedlot eating hay. You have to decide what is acceptable for you—you might choose to only consume animals that were never in a feedlot, or you might be okay with one held in a feedlot for a couple days, or longer if you know how they were treated while there.

The two important issues with feedlots are the density of animals and the type of feed used. If the animals are packed in together, where they spend their time standing in mud and their own feces, this is a factory farm feedlot and should be avoided. If the animals are fed animal by-products, outdated food, grains, and other unnatural products, they are not sustainable.

If you choose to purchase meat from an animal that never spent any time on a feedlot, look for meat that was one hundred percent pasture-raised. If you choose to purchase meat from an animal that spent a minimum amount of time on a feedlot, ask the farmer about the raising and finishing process for the animals, as well as what they were fed and how long the animals were in the feedlot.

4. **Was the animal ever given antibiotics?**

Some consumers only want to eat meat that was never given anti-biotics, even to treat illness. Other consumers are okay with the thera-peutic use of antibiotics, meaning that the animals are treated with antibiotics only if they get sick.

You need to decide which is best for you. Any animal fed a low dose of antibiotics on a continual basis, either to promote growth or to ward off possible disease, is an unsustainable animal that was raised on a factory farm. This type of meat should be avoided.

5. **Were hormones, steroids, or growth promoters ever given to the cow?**

There is only one reason cattle are given hormones, steroids, or any type of growth promoter—to make them grow faster. The practice of giving animals any type of growth promoter is not sustainable and should be avoided, so the answer to this question should be "no."

Another type of hormone is given to dairy cows—rBGH. This genetically engineered hormone is injected for one reason only—to make the animals produce more milk. Cows given rBGH (or rBST) are not sustainable, and their products should be avoided.

## QUESTIONS FOR A FARMER: DAIRY

1. **Was the cow raised on pasture? Or how was the animal raised?**

   When animals are raised on pasture or in fields, they graze outdoors. Cows' stomachs are designed to digest grasses—they are called ruminants and eat grass. You want the farmer to tell you that the cow grazed out on the fields and ate a vegetarian diet.

   Obviously, the animal has to be milked, so some time will probably be spent indoors, but you want to make sure the animal has had adequate time to graze outdoors.

2. **Was the animal fed anything else besides grass?**

   You want to know if the farmer fed the animal any supplements, by-products, or additional types of feed. On factory farms, the animals might be fed such things as cement dust, rotten and outdated food, poultry litter, and other unsavory products.

   Animals are also fed animal by-products, even since the 2003 mad cow disease discovery in the United States. There are efforts to stop the loopholes where cows can be fed back to cows (as of June 2004, the loopholes still existed). But even if all the loopholes are closed, other animals can still be ground up and fed to cows (who, remember, are vegetarians by nature). Supplements often contain animal fat and protein.

   The only way to be sure there are no animal products in the feed being given to the animal is to know that the animal had a one hundred percent vegetarian diet or was given one hundred percent vegetarian feed, though a high amount of corn in the animal's diet can make it sick.

3. **Was the cow given rBGH or any type of synthetic hormone?**

   An estimated thirty percent of all dairy cows are injected with a genetically engineered hormone called recombinant bovine growth hormone (rBGH). This hormone forces cows to produce more milk and often leads to painful udder infections that cause pus and blood to mix in with the milk. Because the animals become more prone to sickness, they are given increased levels of antibiotics and other drugs, and residues of these medications can be found in the milk. Most countries do not allow the use of rBGH because of concerns over the safety of the product. You should not buy milk, cheese, ice cream, yogurt, or any dairy product that was made from cows given rBGH.

4. **Was the animal ever given antibiotics?**

Some consumers want to know the animals were never given antibiotics, even to treat illness. Other consumers are okay with the therapeutic use of antibiotics, meaning that the animals are treated with antibiotics only if they get sick.

You need to decide which is best for you. Any animal fed a low dose of antibiotics on a continual basis, either to promote growth or to ward off possible disease, is an unsustainable animal that was raised on a factory farm. This should be avoided.

5. **How long do dairy cows stay in your herd?**

Sustainable dairy operations keep cows for ten to fifteen years, while operations that use rGBH and other supplements wind up replacing the cows every five to seven years. The high cost this replacement entails is one of the reasons that dairy farmers try to market downer cows (sickly cows that are unable to walk) as meat. And that contributes to mad cow disease.

## QUESTIONS FOR A FARMER: POULTRY

1. **How was the animal raised? On pasture, indoors, confined?**

Studies are starting to show that the animals healthiest for you are those raised on pasture, so you ideally want to find poultry raised outdoors in their natural state.

If you live in a cold part of the country, that might not be feasible, so you want to find poultry that's raised humanely indoors. The chickens should not be overcrowded and should have access to outdoors. What's important are the number of animals raised together, the size of the space they live in, and if they are provided straw and other items they would naturally have access to outdoors.

2. **How much time does the poultry spend outdoors each day?**

There's a big difference between an animal that's permitted access to outdoors for ten minutes a day as opposed to an animal that spends ten hours, or its whole life, outdoors. Ideally, you are looking for an animal that spends a significant amount of time outdoors in the natural environment, though you do need to factor in the part of the country you live in.

In cold or very hot climates, or places where there are cold winters or hot summers, you might need to look at how humanely the animals are raised, and whether they ever get time outdoors. Ideally, they should have continual access to both indoors and outdoors, where they can choose. Factory farm poultry are often raised under artificial lighting that is constantly on, and the birds are packed together very tightly indoors. The farmer should be able to explain to you why s/he raises the animals the way that s/he does—you should feel comfortable with the explanation if the animals are being raised sustainably.

3. **What was the chicken/turkey fed?**

Sustainable poultry eat grasses, greens, grains, and insects, whereas factory farm poultry are fed animal by-products such as bone, feathers, blood, manure, and animal parts, as well as grain, arsenic, mineral and vitamin supplements, enzymes, and antibiotics. If the farmer tells you that the feed was supplemented with anything, dig further to find out exactly what the supplements are.

4. **Was the chicken/turkey given antibiotics?**

Some consumers only want poultry that has never received antibiotics; others are comfortable eating poultry that was treated with antibiotics only when the birds became sick. Any animal given a low dose of antibiotics to promote growth and/or to ward off disease is being raised in a factory farm system and should be avoided.

5. **Were hormones, steroids, or growth promoters ever given to the birds?**

By law, hormones cannot be given to poultry. But animals can be fed growth enhancers and feed additives in order to make the poultry grow faster. These additives are not considered hormones, but there is concern that they might affect human health. It is best to find farmers who do not feed their animals any hormones, growth enhancers, or any type of chemical feed additives.

You also might want to ask if animal protein was fed as an additive or as part of their diet. (Poultry are meat eaters—they eat bugs and insects—so they can be fed animal protein on both sustainable and unsustainable farms.) What you are concerned about is if any of the animal protein fed to poultry contains hormones. If a chicken or turkey is fed beef or a beef by-product, that beef could conceivably contain hormones—this is one way hormones are thought to be getting into the poultry supply. It is uncertain whether this type of hormone transmission is affecting human health, so you must decide whether or not this is important to you.

## QUESTIONS FOR A FARMER: EGGS

1. **How was the animal raised? On pasture, indoors, confined? Was it caged?**

   Studies are starting to show that the animals healthiest for you are those raised on pasture, so you ideally want to find poultry raised outdoors in their natural state.

   If you live in a cold part of the country, that might not be feasible, so you want to find poultry raised humanely indoors. The chickens should not be overcrowded and should have access to outdoors. What's important are the number of animals raised together, the size of the space they live in, and if they are provided straw and other items they would naturally have access to outdoors.

   Egg-laying hens in factory farms spend their life crammed inside cages where they can't even spread their wings. Raising poultry in any type of cage is unsustainable and should be avoided.

2. **How much time do the hens spend outdoors each day?**

   There's a big difference between an animal that's permitted access to outdoors for ten minutes a day as opposed to an animal that spends ten hours, or its whole life, outdoors. Ideally, you are looking for an animal that spends a significant amount of time outdoors in the natural environment, though you do need to factor in the part of the country you live in.

In cold or very hot climates, or places where there are cold winters or hot summers, you might need to look at how humanely the animals are raised, and whether they ever get time outdoors. Ideally, the animals should have continual access to both indoors and outdoors, where they can choose. The farmer should be able to explain to you why s/he raises the animals the way that s/he does—you should feel comfortable with the explanation if the animals are being raised sustainably.

3. **Was the hen force molted?**

Molting is the process where a chicken replaces its old feathers with new ones—it's a natural process that happens every year.

Forced molting is a process where hens are not given any food for between five and fourteen days. This forces their bodies to produce more eggs. Forced molting is inhumane and unsustainable. You should avoid eggs from chickens that were raised this way.

4. **What was the hen fed?**

Sustainable poultry eat grasses, greens, grains, and insects, whereas factory farm poultry are fed animal by-products such as bone, feathers, blood, manure, and animal parts, as well as grain, arsenic, mineral and vitamin supplements, enzymes, and antibiotics. If the farmer tells you that the feed was supplemented with anything, dig further to find out exactly what the supplements are.

5. **Was the hen fed antibiotics?**

Some consumers only want poultry that has never received antibiotics; others are comfortable eating poultry that was treated with antibiotics only when the birds became sick. Any animal given a low dose of antibiotics in order to promote growth and/or ward off disease is being raised in a factory farm system and should be avoided.

Note on hormones: by law, hormones cannot be given to poultry. But animals can be fed growth enhancers and feed additives in order to make the poultry grow faster. These additives are not considered hormones, but there is concern that they might affect human health. It is best to find farmers who do not feed their animals any hormones, growth enhancers, or any type of chemical feed additives.

You also might want to ask if animal protein was fed as an additive or as part of their diet. (Poultry are meat eaters—they eat bugs and insects—so they can be fed animal protein on both sustainable and unsustainable farms.) What you are concerned about is if any of the animal protein fed to poultry contains hormones. If a chicken or turkey is fed beef or a beef by-product, that beef could conceivably contain hormones—this is one way hormones are thought to be getting into the poultry supply. It is uncertain whether this type of hormone transmission is affecting human health, so you must decide whether or not this is important to you.

## QUESTIONS FOR A FARMER: HOGS

1. **How was the hog raised?**

   Ideally, you are looking for a hog that was raised outdoors on pasture and in fields. In some areas of the country, the weather is too cold to permit constant access to outdoors. Pigs in these areas should have comfortable barns or sustainable structures like hoop houses where they have space to carry out their natural behaviors, such as rooting and nesting, and they should be provided proper bedding materials such as straw. They should also have the ability to go outdoors should they wish to.

2. **Where did its mother and father live?**

   One of the most troubling aspects of factory hog production is the treatment of pregnant sows (female pigs) and the boars (males) that breed them. They may spend their entire lives in "farrowing pens" (small crates with metal bars) too small to turn around, standing on slatted floors, with every natural instinct to build nests and nurture piglets thwarted. So even if those piglets are eventually taken to a satisfactory environment, raised "free-range" or the like, you'll want to be sure they weren't born in one of these industrial facilities.

3. **How much time do the animals spend outdoors each day?**

   Having "access" to outdoors isn't good enough—some companies have interpreted that as a small opening out onto a concrete patio. Find out if the hog goes out into the fields or onto pasture, and ask how much time a day the animal spends there. There's a big difference between four minutes and four hours.

4. **What was the hog fed?**

Sustainable hogs root in the dirt and eat roots and bugs, as well as food such as corn, soy, vegetables and vegetable peelings, extra dairy products, and table scraps. An ideal system is where a family farm raises the grain and soybeans fed to the animals. Factory farm hogs are mostly raised on corn and soy, and this can be supplemented with bakery products, limestone, fish meal, copper, choline chloride, antibiotics, blood cells, and sodium selenite, among other things.

5. **Was the hog ever given antibiotics?**

Some consumers only want pork from animals that never received antibiotics; others are comfortable eating pigs that were treated with antibiotics only when the animals became sick. Any animal given a low, daily dose of antibiotics in order to ward off disease is being raised in a factory farm system and should be avoided.

6. **Were hormones or feed additives given to the hogs?**

By law, hormones cannot be given to hogs. But animals can be fed growth enhancers and feed additives in order to make the animals grow faster. These additives are not considered hormones, but there is concern that they might affect human health. It is best to find farmers who do not feed their animals any type of hormones, growth enhancers, or chemical feed additives.

You also might want to ask if animal protein was fed as an additive or as part of their diet. (Hogs are meat eaters, so they can be fed meat on both sustainable and unsustainable farms.) What you are concerned about is if any of the animal protein fed to the hogs contains hormones. If a hog is fed beef or a beef by-product, that beef could conceivably contain hormones—this is one way hormones are thought to be getting into the pork supply. It is uncertain whether this type of hormone transmission is affecting human health, so you must decide whether or not this is important to you.

# ELEVEN

# EATING MADE SIMPLE*

## By Marion Nestle

Marion Nestle is the Paulette Goddard Professor of nutrition, food studies, and public health at New York University, in the department that she chaired from 1988 through 2003. She also holds appointments as professor of sociology in NYU's College of Arts and Sciences and as a visiting professor of nutritional sciences in the College of Agriculture at Cornell University. Her degrees include a doctorate in molecular biology and a master's in public health nutrition, both from the University of California, Berkeley.

Dr. Nestle's research focuses on the politics of food with an emphasis on the role of food marketing as a determinant of dietary choice. She is the author of *Food Politics: How the Food Industry Influences Nutrition and Health* (University of California Press, 2002), *Safe Food: Bacteria, Biotechnology, and Bioterrorism* (University of California Press, 2003), and *What to Eat* (North Point Press, a division of Farrar, Straus and Giroux, 2006) and is coeditor (with Beth Dixon) of *Taking Sides: Clashing Views on Controversial Issues in Food and Nutrition* (McGraw-Hill/Dushkin, 2004). In 2007, *Food Politics* was issued in a revised and expanded edition. Her latest book, *Pet Food Politics: Chihuahua in the Coal Mine*, was published by the University of California Press in September 2008.

A s a nutrition professor, I am constantly asked why nutrition advice seems to change so much and why experts so often disagree. Whose information, people ask, can we trust? I'm tempted to say, "Mine, of course," but I understand the problem. Yes, nutrition advice seems endlessly mired in scientific argument, the self-interest of food companies,

---

and compromises by government regulators. Nevertheless, basic dietary principles are not in dispute: eat less; move more; eat fruits, vegetables, and whole grains; and avoid too much junk food.

"Eat less" means consume fewer calories, which translates into eating smaller portions and steering clear of frequent between-meal snacks. "Move more" refers to the need to balance calorie intake with physical activity. Eating fruits, vegetables, and whole grains provides nutrients unavailable from other foods. Avoiding junk food means to shun "foods of minimal nutritional value"—highly processed sweets and snacks laden with salt, sugars, and artificial additives. Soft drinks are the prototypical junk food; they contain sweeteners but few or no nutrients.

If you follow these precepts, other aspects of the diet matter much less. Ironically, this advice has not changed in years. The noted cardiologist Ancel Keys (who died in 2004 at the age of 100) and his wife, Margaret, suggested similar principles for preventing coronary heart disease nearly fifty years ago.

But I can see why dietary advice seems like a moving target. Nutrition research is so difficult to conduct that it seldom produces unambiguous results. Ambiguity requires interpretation. And interpretation is influenced by the individual's point of view, which can become thoroughly entangled with the science.

## NUTRITION SCIENCE CHALLENGES

This scientific uncertainty is not overly surprising given that humans eat so many different foods. For any individual, the health effects of diets are modulated by genetics but also by education and income levels, job satisfaction, physical fitness, and the use of cigarettes or alcohol. To simplify this situation, researchers typically examine the effects of single dietary components one by one.

Studies focusing on one nutrient in isolation have worked splendidly to explain symptoms caused by deficiencies of vitamins or minerals. But this approach is less useful for chronic conditions, such as coronary heart disease and diabetes, that are caused by the interaction of dietary, genetic, behavioral, and social factors. If nutrition science seems puzzling, it is because researchers typically examine single nutrients detached from food itself, foods separate from diets, and risk factors apart from other behaviors. This kind of research is "reductive" in that it attributes health effects to the consumption of one nutrient or food when it is the overall dietary pattern that really counts most.

For chronic diseases, single nutrients usually alter risk by amounts too small to measure except through large, costly population studies. As seen recently in the Women's Health Initiative, a clinical trial that examined the effects of low-fat diets on heart disease and cancer, participants were unable to stick with the restrictive dietary protocols. Because humans cannot be caged and fed measured formulas, the diets of experimental and control study groups tend to converge, making differences indistinguishable over the long run—even with fancy statistics.

## IT'S THE CALORIES

Food companies prefer studies of single nutrients because they can use the results to sell products. Add vitamins to candies, and you can market them as health foods. Health claims on the labels of junk foods distract consumers from their caloric content. This practice matters because when it comes to obesity—which dominates nutrition problems even in some of the poorest countries of the world—it is the calories that count. Obesity arises when people consume significantly more calories than they expend in physical activity.

America's obesity rates began to rise sharply in the early 1980s. Sociologists often attribute the "calories in" side of this trend to the demands of an overworked population for convenience foods—prepared, packaged products and restaurant meals that usually contain more calories than home-cooked meals.

But other social forces also promoted the calorie imbalance. The arrival of the Reagan administration in 1980 increased the pace of industry deregulation, removing controls on agricultural production and encouraging farmers to grow more food. Calories available per capita in the national food supply (that produced by American farmers, plus imports, less exports) rose from 3,200 a day in 1980 to 3,900 a day two decades later.

The early 1980s also marked the advent of the "shareholder value movement" on Wall Street. Stockholder demands for higher short-term returns on investments forced food companies to expand sales in a marketplace that already contained excessive calories. Food companies responded by seeking new sales and marketing opportunities. They encouraged formerly shunned practices that eventually changed social norms, such as frequent between-meal snacking, eating in book and clothing stores, and serving larger portions. The industry continued to sponsor organizations and journals that focus on nutrition-related subjects and intensified its efforts to lobby government for favorable dietary advice. Then and now, food lobbies have promoted positive interpretations of scientific studies, sponsored research that can be used as a basis for health claims, and attacked critics, myself among them, as proponents of "junk science." If anything, such activities only add to public confusion.

## SUPERMARKETS AS "GROUND ZERO"

No matter whom I speak to, I hear pleas for help in dealing with supermarkets, considered by shoppers as "ground zero" for distinguishing health claims from scientific advice. So I spent a year visiting supermarkets to help people think more clearly about food choices. The result was my book *What to Eat*.

Supermarkets provide a vital public service but are not social services agencies. Their job is to sell as much food as possible. Every aspect of store design—from shelf position to background music—is based on marketing research. Because this research shows that the more products customers see, the more they buy, a store's objective is to expose shoppers to the maximum number of products they will tolerate viewing.

If consumers are confused about which foods to buy, it is surely because the choices require knowledge of issues that are not easily resolved by science and are strongly swayed by social and economic considerations. Such decisions play out every day in every store aisle.

## ARE ORGANICS HEALTHIER?

Organic foods are the fastest-growing segment of the industry, in part because people are willing to pay more for foods that they believe are healthier and more nutritious. The U.S. Department of Agriculture forbids producers of "Certified Organic" fruits and vegetables from using synthetic pesticides, herbicides, fertilizers, genetically modified seeds, irradiation, or fertilizer derived from sewage sludge. It licenses inspectors to ensure that producers follow those rules. Although the USDA is responsible for organics, its principal mandate is to promote conventional agriculture, which explains why the department asserts that it "makes no claims that organically produced food is safer or more nutritious than conventionally produced food. Organic food differs from conventionally grown food in the way it is grown, handled and processed."

This statement implies that such differences are unimportant. Critics of organic foods would agree; they question the reliability of organic certification and the productivity, safety, and health benefits of organic production methods. Meanwhile the organic food industry longs for research to address such criticisms, but studies are expensive and difficult to conduct. Nevertheless, existing research in this area has established that organic farms are nearly as productive as conventional farms, use less energy, and leave soils in better condition. People who eat foods grown without synthetic pesticides ought to have fewer such chemicals in their bodies, and they do. Because the organic rules require pretreatment of manure and other steps to reduce the amount of pathogens in soil treatments, organic foods should be just as safe—or safer—than conventional foods.

Similarly, organic foods ought to be at least as nutritious as conventional foods. And proving organics to be more nutritious could help justify their higher prices. For minerals, this task is not difficult. The mineral content of plants depends on the amounts present in the soil in which they are grown. Organic foods are cultivated in richer soils, so their mineral content is higher.

But differences are harder to demonstrate for vitamins or antioxidants (plant substances that reduce tissue damage induced by free radicals); higher levels of these nutrients relate more to a food plant's genetic strain or protection from unfavorable conditions after harvesting than to production methods. Still, preliminary studies show benefits: organic peaches and pears contain greater quantities of vitamins C and E, and organic berries and corn contain more antioxidants.

Further research will likely confirm that organic foods contain higher nutrient levels, but it is unclear whether these nutrients would make a measurable improvement in health. All fruits and vegetables contain useful nutrients, albeit in different combinations and concentrations. Eating a variety of food plants is surely more important to health than small differences in the nutrient content of any one food. Organics may be somewhat healthier to eat, but they are far less likely to damage the environment, and that is reason enough to choose them at the supermarket.

## DAIRY AND CALCIUM

Scientists cannot easily resolve questions about the health effects of dairy foods. Milk has many components, and the health of people who consume milk or dairy foods is influenced by everything else they eat and do. But this area of research is especially controversial because it affects an industry that vigorously promotes dairy products as beneficial and opposes suggestions to the contrary.

Dairy foods contribute about seventy percent of the calcium in American diets. This necessary mineral is a principal constituent of bones, which constantly lose and regain calcium during normal metabolism. Diets must contain enough calcium to replace losses or else bones become prone to fracture. Experts advise consumption of at least one gram of calcium a day to replace everyday losses. Only dairy foods provide this much calcium without supplementation.

But bones are not just made of calcium; they require the full complement of essential nutrients to maintain strength. Bones are stronger in people who are physically active and who do not smoke cigarettes or drink much alcohol. Studies examining the effects of single nutrients in dairy foods show that some nutritional factors—magnesium, potassium, vitamin D, and lactose, for example—promote calcium retention in bones. Others, such as protein, phosphorus, and sodium, foster calcium excretion. So bone strength depends more on overall patterns of diet and behavior than simply on calcium intake.

Populations that do not typically consume dairy products appear to exhibit lower rates of bone fracture despite consuming far less calcium than recommended. Why this is so is unclear. Perhaps their diets contain less protein from meat and dairy foods, less sodium from processed foods, and less phosphorus from soft drinks, so they retain calcium more effectively. The fact that calcium balance depends on multiple factors could explain why rates of osteoporosis (bone density loss) are highest in countries where people eat the most dairy foods. Further research may clarify such counterintuitive observations.

In the meantime, dairy foods are fine to eat if you like them, but they are not a nutritional requirement. Think of cows: they do not drink milk after weaning, but their bones support bodies weighing 800 pounds or more. Cows feed on grass, and grass contains calcium in small amounts—but those amounts add up. If you eat plenty of fruits, vegetables, and whole grains, you can have healthy bones without having to consume dairy foods.

## A MEATY DEBATE

Critics point to meat as the culprit responsible for elevating blood cholesterol, along with raising risks for heart disease, cancer, and other conditions. Supporters cite the lack of compelling science to justify such allegations; they emphasize the nutritional benefits of meat protein, vitamins, and minerals. Indeed, studies in developing countries demonstrate health improvements when growing children are fed even small amounts of meat.

But because bacteria in a cow's rumen attach hydrogen atoms to unsaturated fatty acids, beef fat is highly saturated—the kind of fat that increases the risk of coronary heart disease. All fats and oils contain some saturated fatty acids, but animal fats, especially those from beef, have more saturated fatty acids than vegetable fats. Nutritionists recommend eating no more than a heaping tablespoon (twenty grams) of saturated fatty acids a day. Beef eaters easily meet or exceed this limit. The smallest McDonald's cheeseburger contains six grams of saturated fatty acids, but a Hardee's Monster Thickburger has forty-five grams.

Why meat might boost cancer risks, however, is a matter of speculation. Scientists began to link meat to cancer in the 1970s, but even after decades of subsequent research, they remain unsure if the relevant factor might be fat, saturated fat, protein, carcinogens, or something else related to meat. By the late 1990s, experts could conclude only that eating beef probably increases the risk of colon and rectal cancers and possibly enhances the odds of acquiring breast, prostate, and perhaps other cancers. Faced with this uncertainty, the American Cancer Society suggests selecting leaner cuts, smaller portions, and alternatives such as chicken, fish, or beans— steps consistent with today's basic advice about what to eat.

## FISH AND HEART DISEASE

Fatty fish are the most important sources of long-chain omega-3 fatty acids. In the early 1970s, Danish investigators observed surprisingly low frequencies of heart disease among indigenous populations in Greenland that typically ate fatty fish, seals, and whales. The researchers attributed the protective effect to the foods' content of omega-3 fatty acids. Some subsequent studies—but by no means all—confirm this idea.

Because large, fatty fish are likely to have accumulated methyl mercury and other toxins through predation, however, eating them raises questions about the balance between benefits and risks. Understandably, the fish industry is eager to prove that the health benefits of omega-3s outweigh any risks from eating fish.

Even independent studies on omega-3 fats can be interpreted differently. In 2004, the National Oceanic and Atmospheric Administration— for fish, the agency equivalent to the USDA—asked the Institute of Medicine (IOM) to review studies of the benefits and risks of consuming seafood. The ensuing review of the research on heart disease risk illustrates the challenge such work poses for interpretation.

The IOM's October 2006 report concluded that eating seafood reduces the risk of heart disease but judged the studies too inconsistent to decide if omega-3 fats were responsible. In contrast, investigators from the Harvard School of Public Health published a much more positive report in the *Journal of the American Medical Association* that same month. Even modest consumption of fish omega-3s, they stated, would cut coronary deaths by thirty-six percent and total mortality by seventeen percent, meaning that not eating fish would constitute a health risk.

Differences in interpretation explain how distinguished scientists could arrive at such different conclusions after considering the same studies. The two groups, for example, had conflicting views of earlier work published in March 2006 in the *British Medical Journal*. That study found no overall effect of omega-3s on heart disease risk or mortality, although a subset of the original studies displayed a fourteen percent reduction in total mortality that did not reach statistical significance. The IOM team interpreted the "nonsignificant" result as evidence for the need for caution, whereas the Harvard group saw the data as consistent with studies reporting the benefits of omega-3s. When studies present inconsistent results, both interpretations are plausible. I favor caution in such situations, but not everyone agrees.

Because findings are inconsistent, so is dietary advice about eating fish. The American Heart Association recommends that adults eat fatty fish at least twice a week, but U.S. dietary guidelines say: "Limited evidence suggests an association between consumption of fatty acids in fish and reduced risks of mortality from cardiovascular disease for the general population . . . however, more research is needed." Whether or not fish uniquely protects against heart disease, seafood is a delicious source of many nutrients, and two small servings per week of the less predatory classes of fish are unlikely to cause harm.

## SODAS AND OBESITY

Sugars and corn sweeteners account for a large fraction of the calories in many supermarket foods, and virtually all the calories in drinks—soft, sports, and juice—come from added sugars.

In a trend that correlates closely with rising rates of obesity, daily per capita consumption of sweetened beverages has grown by about 200 calories since the early 1980s. Although common sense suggests that this increase might have something to do with weight gain, beverage makers argue that studies cannot prove that sugary drinks alone—independent of calories or other foods in the diet—boost the risk of obesity. The evidence, they say correctly, is circumstantial. But pediatricians often see obese children in their practices who consume more than 1,000 calories a day from sweetened drinks alone, and several studies indicate that children who habitually consume sugary beverages take in more calories and weigh more than those who do not.

Nevertheless, the effects of sweetened drinks on obesity continue to be subject to interpretation. In 2006, for example, a systematic review funded by independent sources found sweetened drinks promoted obesity in both children and adults. But a review that same year sponsored in part by a beverage trade association concluded that soft drinks have no special role in obesity. The industry-funded researchers criticized existing studies as being short-term and inconclusive and pointed to studies finding that people lose weight when they substitute sweetened drinks for their usual meals.

These differences imply the need to scrutinize food industry sponsorship of research itself. Although many researchers are offended by suggestions that funding support might affect the way they design or interpret studies, systematic analyses say otherwise. In 2007, investigators classified studies of the effects of sweetened and other beverages on health according to who had sponsored them. Industry-supported studies were more likely to yield results favorable to the sponsor than those funded by independent sources. Even though scientists may not be able to prove that sweetened drinks cause obesity, it makes sense for anyone interested in losing weight to consume less of them.

The examples I have discussed illustrate why nutrition science seems so controversial. Without improved methods to ensure compliance with dietary regimens, research debates are likely to rage unabated. Opposing points of view and the focus of studies and food advertising on single nutrients rather than on dietary patterns continue to fuel these disputes. While we wait for investigators to find better ways to study nutrition and health, my approach—eat less, move more, eat a largely plant-based diet, and avoid eating too much junk food—makes sense and leaves you plenty of opportunity to enjoy your dinner.

# Another Take | WORLD HUNGER—YOUR ACTIONS MATTER

## By Sherri White Nelson, Heifer International

---

The mission of Heifer International is to end hunger and poverty while caring for the Earth. For more than sixty years, the organization has provided livestock and environmentally sound agricultural training to improve the lives of those who struggle daily for reliable sources of food and income, helping forty-eight million people through training in livestock development and livestock gifts that multiply.

Every gift of an animal through Heifer International provides direct benefits, such as milk, eggs, wool, and fertilizer, as well as indirect benefits that increase family incomes for better housing, nutrition, health care, and school fees for children. Recipients "pass on the gift" of offspring of their cows, goats, and other livestock to others in an ever-widening circle of hope. Heifer currently works in more than fifty-seven countries, including the United States.

In "World Hunger—Your Actions Matter," Heifer staffer Sherri White Nelson provides simple answers to a question many people ask: what can I do, starting today, to help alleviate the problem of world hunger? For more information, visit Heifer International's website at http://www.heifer.org.

---

A shy mother of three, Josefa Hernandez has a broad, infectious smile, but her brown, weathered face and worry-lined eyes reveal the hardships she endures. I met Josefa in the summer of 2005 in a remote mountain village near Cuetzalan, Mexico, while on assignment for Heifer International. Of all the struggling people I have met in the course of my job as Heifer's magazine editor, Josefa is the one I think of most.

Josefa wakes every day at 4:00 a.m. to start the morning fire in a large tin pail at the corner of her one-room, dirt-floored home. Black smoke fills the air and seeps out into the break-of-day fog through open gaps in the shoddy wooden planks that pass for walls. A huge portion of the front wall has no planks at all—only a torn plastic tarp tied by rope to the adjoining lumber.

Each morning she heats water for her children's baths, which take place outside because there is no bathroom, no shower, no toilet. She has no electricity or running water. After her son and youngest daughter go to school, Josefa begins her journey into the forest to fetch firewood before the tropical heat rises to unbearable temperatures. She straps her disabled middle child, Andrea, on her back with a *rebozo*, a handwoven scarf, and begins the laborious journey into the woods.

I walk with Josefa for what seems like miles, up and down steep hillsides on dangerous paths littered with ankle-breaking holes left from planting coffee crops. We both gasp for breath in the thick, humid air as sweat streams down our faces. She carefully chooses the limbs that are safe to take and hesitantly asks if I can carry a load of wood with her. The coffee plantation owner has strict rules about which branches she's allowed to use. With heavy stacks of wood, we begin the long walk to her home. I struggle to carry the weight as she adeptly maneuvers the trails, wood in hand, child on back. Josefa amazes me, and I feel very small in her presence.

After we unload the firewood, Josefa and I sit on the ground, and I ask her how she manages to do this several times a week, alone with a sick child. Andrea, then fourteen, weighed less than forty pounds. She will never speak or walk or care for herself. With a mixture of pride and sorrow, her mother says, "Andrea tickles me, hugs me, and gives me kisses. But I would love for her to be able to talk to me and walk on her own. But I know that's not going to happen."

Josefa and her family live and work on land owned by a coffee plantation owner. She has no rights to the property and makes a paltry wage. She struggles to afford her children's school fees and medical care. Josefa is at once the face of hope and the face of despair, of loss and of promise. A simple donation of chickens from Heifer International gives her a hand up toward providing for her family, to turn her hard work into something tangible: eggs for nutrition, income for necessities.

## THE FACES OF HUNGER

It is hard to comprehend that the number of undernourished people rose to 963 million in 2008.[1] One billion people live in extreme poverty, living on less than \$1 per day, and half the world's population lives on less than \$2.[2] Even in the United States, more than thirty-six million are hungry.[3] And with the worsening global food and economic crises, these numbers are sure to rise.

Generous people who support charities have less to donate. Corporations are reducing or completely eliminating their philanthropic programs. As unemployment rises, many once self-sufficient families must depend on charities to meet basic needs for food, shelter, and clothing. Yet these charities now have more people to serve with much less to give. A dangerous cycle emerges. The problem seems so insurmountable that it's easy to feel helpless and hopeless—and that is the biggest danger of all. It is still possible to end hunger; I've found it begins to make sense if you look at it by first taking one action, helping one family in need.

We have tough choices to make. Even with our shrinking wallets and busy lives, we still have the power to make a difference for people in need, globally and locally. Often, the price tag is simply action and commitment. We can quiet hunger for Josefa and others who live a world away, and for our neighbors who live just down the street. And we can do it stress-free and in many cases, money-free.

Be a world-changer. Get involved. Make a difference. Your actions matter. There's a great feel-good factor when you commit to a cause you believe in. I've seen the results firsthand. For inspiration, review these simple suggestions and follow your heart; it will tell you where your contributions will have the greatest effect.

## TEN EASY WAYS TO HELP END HUNGER

1. **Speak with ONE voice.** Never have time to write that letter to your congressperson expressing your views on global issues? To increase your influence on issues of poverty and hunger, tap into the power of ONE.org. ONE is a nonpartisan campaign of more than 2.4 million

people from all fifty states and more than one hundred of the United States' most effective and respected nonprofit, advocacy, and humanitarian organizations. Partners include Bread for the World, UNICEF, Habitat for Humanity, Heifer International, CARE USA, and Mercy Corps. ONE aims to raise public awareness about the issues of global poverty, hunger, and disease and encourages efforts to fight such problems in the world's poorest countries.

The organization asks our leaders to do more to fight global AIDS and extreme poverty. Working in communities, colleges, and churches across America, ONE members educate and ask politicians from every level—from city councils to the U.S. Congress—to approve legislation to increase the effectiveness of U.S. and international assistance, to make trade fair, and to fight corruption. The website is easy to use, and they do all the hard work and research for you. Plus, you get a nice sense of community when you see all the people—just like you and me—combining the power of our voices and views to help end poverty. Lend your voice at www.one.org.

2. **Play a game.** Sound too good to be true? Playing a fun vocabulary game online actually helps end hunger. FreeRice will donate twenty grains of rice for each question you answer correctly. Twenty grains might not sound like a lot, but the organization, started in October 2007, has donated more than forty billion grains of rice and has fed more than one million people. Put your word skills to the test and task FreeRice with helping a million more people. Check it out at www.freerice.com.

3. **Bake a cake.** Here's to a cause that satisfies your inner humanitarian and your sweet tooth. Share Our Strength's Great American Bake Sale® works to end childhood hunger by recruiting volunteers to host bake sales in their communities. The proceeds go to community organizations across America to provide at-risk children with healthful meals when most needed: during the summer and after school. Since 2003, more than one million people have participated in the Great American Bake Sale, raising nearly $4 million to make sure no child in America grows up hungry. Put your baking skills to work for hungry kids. Go to http.gabs.strength.org to learn how you can get involved.

4. **Unwrap change.** If finding the perfect gift for you is mission impossible, make life easier for your family and friends by asking for

donations to your favorite charities. Many nonprofits like Heifer now have gift registries online, and you can use them much like the wish lists found at Amazon.com. Any special occasion—birthdays, Christmas, Hanukkah, bar and bat mitzvahs, graduations, weddings—can bring awareness and needed funds to a worthy cause. It's the one gift you'll never have to exchange.

5. **Shop with a purpose.** Things don't simply appear on store shelves out of nowhere. They have all been planted and picked, cut and sewn, carved and sanded by real people. While you may never see them, you can ensure their working lives are as fair as possible by buying fair-trade certified foods, non-sweatshop textiles and clothing, and products from companies that promise to pay workers a living wage. It's important to remember the "invisible" people on the other end of our purchases.

   Global Goods Partners, www.globalgoodspartners.org, can help you make your next shopping spree one that helps end poverty and promotes social justice. This nonprofit strengthens women-led development initiatives and creates access to the U.S. market for marginalized communities in Asia, Africa, and the Americas. Because Global Goods Partners is a fair-trade organization, women workers in poor communities earn a fair income to support themselves and their families.

6. **Leave a smaller footprint.** You may not think of hunger when you think of climate change. But in countries like India and Tanzania, rising temperatures are causing the glaciers to recede. These glaciers feed the rivers, which provide drinking water and water for crops. Disappearing glaciers equal disappearing water supplies. So to positively impact climate change and protect the poor, do what you can to reduce your carbon footprint by cutting unnecessary use of finite resources.

   There are many ways, big and small, to protect the environment. Turn off the water when you brush your teeth. Fix leaky faucets. Ditch the plastic bags and take reusable bags on your next shopping trip. You can reduce your home energy use by ten percent or more by unplugging electronic devices such as game consoles, computer printers, and DVD players when not in use; adding weather stripping to doors and windows to increase the efficiency of heating and air conditioning units; and lowering the temperature of your water heater or using a thermal cover to keep heat in.

The We Campaign offers many commonsense ways to reduce your carbon footprint at www.wecansolveit.org. Here, you'll find everyday actions to take that will benefit the world—and add green to your wallet. For example, turn down the heat and air conditioning when you aren't home. Try using a programmable thermostat or setting your thermostat yourself to 68 degrees while you are awake and lower it to 60 degrees while you are asleep or away from home. In the summer, keep the thermostat at 78 degrees while you are at home, but give your air conditioning a rest when you are away. This will allow you to save about ten percent a year on your home energy costs. If every house in America did this, our total greenhouse gas production would drop by about thirty-five million tons of carbon dioxide. This is about the same as taking six million cars off of the road.

7. **Share the leftovers.** So, you never quite mastered the art of cooking just enough food to feed your family, and you feel a pang of guilt every time you drag an overflowing garbage bag of food to the trash. Take solace in knowing that your leftovers can fight hunger in your very own community.

The news is filled with media reports about how Meals on Wheels can't keep up with increased demand and decreased donations. Food banks and shelters across the nation have bare shelves. 1.3 million more Americans enrolled in the federal Food Stamp Program in 2008 compared to the previous year.[4]

Package your family's leftovers and take them to an elderly person in your church or neighborhood. The current recession, rising food and medical costs, and fixed monthly incomes often leave senior citizens with a choice between food and medicine. What you were going to throw out will provide nourishment and a hot, tasty meal for someone in need. Bonus feel-good factor: you'll probably strike up a good friendship and brighten someone's day.

And if you're the type of person who enjoys this hands-on community approach, try these easy ways to bolster supplies for food banks and shelters:

8. **Plant a row.** When planting your garden this year, plant an extra row or two and donate excess produce to a local food bank or soup kitchen. Plant A Row (PAR) is a public-service campaign led by the Garden Writer's Association, a nonprofit based in Virginia. PAR donations have reached nearly twelve million pounds of food. Learn how you can get involved at www.gardenwriters.org/par. You'll find how-to information, communication supplies, and starter kits. And if you need help finding food banks in your area, go to www.secondharvest.org.

9. **Do more for less.** Grocery stores and bulk supermarkets are filled with unbelievable savings, but what in the world can you do with twenty cans of green beans for $10? Take advantage of the coupons and bulk packaging deals. Take what your family needs, and then donate the rest to your local food bank, shelter, or soup kitchen. You can make this a weekly or monthly trip. It's also a great way to show your children there are many ways, big and small, to lend a helping hand.

10. **Feed your mind.** If you need some inspiration to get you excited about how you're going to make a real and lasting difference, start by curling up with a good read. (This, too, counts as an action.) These are my favorite books about people who change the world and the global issues of hunger and poverty.

Tracy Kidder, *Mountains Beyond Mountains: The Quest of Dr. Paul Farmer, a Man Who Would Cure the World*. New York: Random House, 2004.

Alephonsion Deng, Benson Deng, and Benjamin Ajak, *They Poured Fire on Us from the Sky: The True Story of Three Lost Boys from Sudan*. New York: PublicAffairs, 2006.

Peter Menzel and Faith D'Aluisio, *Hungry Planet: What the World Eats*. Berkeley, CA: Ten Speed Press, 2007.

Wangari Maathai, *Unbowed*. New York: Vintage, 2007.

David K. Shipler, *The Working Poor, Invisible in America*. New York: Vintage, 2005.

Sharman Apt Russell, *Hunger: An Unnatural History*. New York: Basic Books, 2006.

# TWELVE

# IMPROVING KIDS' NUTRITION:
## AN ACTION TOOL KIT FOR PARENTS AND CITIZENS

### By the Center for Science in the Public Interest

Since 1971, the Center for Science in the Public Interest (CSPI) has been a strong advocate for nutrition and health, food safety, alcohol policy, and sound science. Its award-winning newsletter, *Nutrition Action Healthletter*, with some 900,000 subscribers in the United States and Canada, is the largest-circulation health newsletter in North America.

Founded by executive director Michael Jacobson, PhD, and two other scientists, CSPI carved out a niche as the organized voice of the American public on nutrition, food safety, health, and other issues during a boom of consumer and environmental protection awareness in the early 1970s. CSPI has long sought to educate the public, advocate government policies that are consistent with scientific evidence on health and environmental issues, and counter industry's powerful influence on public opinion and public policies.

The information and ideas in this chapter are adapted from materials published by CSPI and available on the organization's website. For more information and regular updates on issues and activities related to nutrition, health, and related problems, visit the site at http://www.cspinet.org.

For millions of Americans, becoming parents brings home the importance of food safety, quality, and nutrition with special force. As parents, we often find ourselves caring more than ever about what and how we eat—partly because taking care of our own health is more important now that we have other lives depending upon us, but especially because we want to raise kids who will be healthy, strong, energetic, and long-lived.

The American system for producing, distributing, and marketing food creates huge challenges for parents who want to instill healthy eating habits in their families. Fortunately, there are plenty of things we can do, both as individuals caring for our own children and as citizens working to improve the quality of nutrition for everyone in our society.

In this chapter, the Center for Science in the Public Interest presents a tool kit for families, providing facts, information, guidelines, and action plans for parents who want to make a difference in the way their kids— and all of America's kids—are eating.

## THE PROBLEM: CHILDHOOD OBESITY
## AND OTHER FOOD-RELATED HEALTH PROBLEMS

As parents, we all want the best for our kids. Most parents do a conscientious job of scheduling medical checkups, getting vaccinations, and taking care of sicknesses that crop up. But many fail to realize that the United States is suffering an epidemic of childhood health problems due to poor nutrition. Consider the following statistics:

- Heart disease, cancer, stroke, and diabetes are responsible for two-thirds of deaths in the United States.[1] The major risk factors for those diseases often are established in childhood: unhealthy eating habits, physical inactivity, obesity, and tobacco use.
- One quarter of children ages five to ten years show early warning signs for heart disease, such as elevated blood cholesterol or high blood pressure.[2]
- Atherosclerosis (clogged arteries) begins in childhood. Autopsy studies of fifteen- to nineteen-year-olds have found that all have fatty streaks in more than one artery, and about ten percent have advanced fibrous plaques.[3]
- Type 2 diabetes can no longer be called "adult onset" diabetes because of rising rates in children. In a study conducted in Cincinnati, the incidence of type 2 diabetes in adolescents increased tenfold between 1982 and 1994.[4] As the number of young people with type 2 diabetes increases, diabetic complications like limb amputations, blindness,

kidney failure, and heart disease will develop at younger ages (likely in their thirties and forties).

- Obesity rates have doubled in children and tripled in adolescents over the last two decades. One in seven young people is obese and one in three is overweight.[5] Obese children are twice as likely as non-obese children to become obese adults.[6]

- Obesity increases the risk of high blood cholesterol, high blood pressure, and diabetes while still in childhood. Excess weight and obesity can result in negative social consequences, e.g., discrimination, depression, and lower self-esteem.[7]

- From 1979 to 1999, annual hospital costs for treating obesity-related diseases in children rose threefold (from $35 million to $127 million).[8]

Further evidence shows the link between these disturbing trends and the bad food choices children learn to make in the early years of their lives. Unhealthy eating habits often begin in childhood.

- Between 1989 and 1996, children's calorie intake increased by approximately 80 to 230 extra calories per day depending on the child's age and activity level.[9] The increases in calorie intake are driven by increased intakes of foods and beverages high in added sugars.

- Only two percent of children (two to nineteen years) meet the recommendations for a healthy diet from the Food Guide Pyramid.[10]

- Three out of four American high school students do not eat the recommended five or more servings of fruits and vegetables each day.[11] Three out of four children consume more saturated fat than is recommended in the *Dietary Guidelines for Americans*.[12]

- Soft drink consumption doubled over the last thirty years.[13] With each additional serving of soft drink (soda, juice drinks, etc.) consumed each day, the odds that a child will become obese increase by sixty percent.[14] Consumption of soft drinks can displace low-fat milk and one hundred percent juice from children's diets.[15] Between 1976 and 1978, boys consumed twice as much milk as soft drinks, and girls consumed fifty percent more milk than soft drinks. By 1994–1996, both boys and girls consumed twice as much soda pop as milk.[16]

Responding to this crisis in children's nutrition requires a multi-pronged approach. In this chapter, we'll discuss several areas parents can focus on: improving the quality of foods provided to kids in schools; advocating for more responsible marketing of foods to children; calling for information efforts, such as nutrition labeling in fast-food restaurants; and, of course, making better food choices for your own kids.

## SCHOOL FOODS—A CRUCIAL FRONT IN THE BATTLE FOR QUALITY KIDS' NUTRITION

School-aged kids eat many of their meals away from home—and for millions, that means eating at school. Unfortunately, many parents have no idea what their children are served at school, and many feel powerless to influence the choices made there. Yet all too often, school foods undermine parents' efforts to feed their children healthfully. Many school leaders simply don't know much about nutrition. Others feel compelled by budget constraints to rely on commercial providers of fast or junk food to feed school kids. And still others assume that children aren't willing to eat healthy foods and therefore take the path of least resistance, offering kids popular foods that may have a devastating long-term effect on their eating habits and health status.

To help parents understand how and what kids may be eating at school, we'll begin with an overview of the food programs most American schools participate in.

The Child Nutrition Programs constitute the federal government's approach to addressing the issue of children's access to nutritious food and nutrition education. The U.S. Department of Agriculture (USDA) is the agency with primary jurisdiction over these programs.

### The National School Lunch Program

The National School Lunch Program (NSLP) was created in 1946 "as a measure of national security, to safeguard the health and well-being of the Nation's children and to encourage the domestic consumption of nutritious agricultural commodities and other food."[17] The NSLP is administered by the USDA and provides meals in 85,000 public (93 percent of the total), 6,500 private schools, and 6,000 child-care institutions. More than fifty percent of U.S. children obtain either breakfast or lunch from the school meal programs.[18] The NSLP serves more than thirty million children daily.[19]

School lunches must meet nutrition standards in order for a school food service program to receive federal subsidies. School lunches must contain less than thirty percent of calories from fat and less than ten percent of calories from saturated fat. In addition, school lunches must provide one-third of the recommended dietary allowances (RDA) for protein, calcium, iron, vitamins A and C, and calories. Legislative language makes it harder for schools to serve only low-fat milk to students.

### The School Breakfast Program

The School Breakfast Program (SBP) was created as a grant program in 1966 to serve breakfasts to "nutritionally needy" children. In 1975, the SBP was permanently authorized as an entitlement program. The SBP serves more than 9.7 million children daily, and fifty-seven percent of schools and seventy-one percent of public schools participate in the program.[20]

School breakfasts must meet nutrition standards in order for a school food service program to receive federal subsidies. School breakfasts must contain less than thirty percent of calories from fat and less than ten percent of calories from saturated fat. In addition, school breakfasts must provide one-fourth of the RDA for protein, calcium, iron, vitamins A and C, and calories.

### After-School Snacks

In 1998, the USDA began providing reimbursements for snacks served to children in after-school programs that meet certain curriculum and student income criteria. The goals of providing snacks are to give children a nutritional "boost" and enable them to take full advantage of educational and enrichment opportunities offered in after-school programs. The after-school snack program provides snacks to 850,000 children.

The budget for the after-school snack program is part of the National School Lunch Program and Child and Adult Care Food Program budgets. In order to receive federal reimbursements, after-school snacks must include a serving of two components from among the following four categories: vegetable, fruit or one hundred percent vegetable or fruit juice; fluid milk; meat or meat alternative; or whole grain, enriched bread, and/or cereal.

### The Team Nutrition Program

The USDA's Team Nutrition program provides nutrition education materials to schools for children and their families. In addition, Team Nutrition uses classroom activities, school-wide events, community programs, and the media to promote healthy eating to children. The program also assists school food service directors with improving the nutritional quality of school meals.

Currently fifty million school children in more than 96,000 schools nationwide participate in Team Nutrition.

### The Special Milk Program

The Special Milk Program provides milk to children who do not receive it through the other USDA Child Nutrition Programs, including the NSLP and SBP. In 2007, 4,914 schools and residential child-care institutions participated in the Special Milk Program, as did 853 summer camps and 533 nonresidential child-care institutions.

Currently, forty-four percent of the milk ordered by schools is high in fat—either two percent or whole milk.[21] Ninety-seven percent of schools offer low-fat or fat-free milk. However, thirty-one percent still offer whole milk.[22] Milk is by far the largest source of saturated fat—the kind of fat that causes heart disease—in children's diets, providing one-quarter of their intake.[23] Because ninety-five percent of maximum bone density is reached by age eighteen, it is especially important that children consume enough calcium to prevent future osteoporosis. Low-fat and fat-free milk provide the calcium children need without the saturated fat that can cause heart disease later in life.

## SCHOOL FOOD INCLUDES
## MORE THAN JUST MEALS:
## WHAT ARE COMPETITIVE FOODS?

"Competitive" foods are foods and beverages served in schools that are not part of USDA school meal programs (i.e., school breakfast, school lunch, or after-school snack programs). They are served or sold in a variety of campus venues, such as à la carte in the cafeteria, vending machines, school stores, fundraisers, and snack bars. Competitive foods are increasingly available to students. Nationally, eighty-three percent of elementary schools, ninety-seven percent of middle/junior high schools, and ninety-nine percent of senior high schools sell foods and beverages out of vending machines, school stores, or à la carte in the cafeteria.[24] The availability of vending machines in schools has increased since the early 1990s. Between 1991 and 2005, the percentage of middle schools with vending machines increased from forty-two percent to eighty-two percent and the percentage of high schools with vending machines increased from seventy-six percent to ninety-seven.[25]

While school meals are required to meet comprehensive nutrition standards, competitive foods are not required to meet those standards. Currently, the USDA has limited authority to regulate the nutritional quality of competitive foods. During meal periods, the sale of "foods of minimal nutritional value" (FMNV) is prohibited by federal regulations in areas of the school where USDA school meals are sold or eaten. However, FMNV can be sold anywhere else on-campus—including just outside the cafeteria—at any time.

An FMNV provides less than five percent of the reference daily intake (RDI) for each of eight specified nutrients per serving.[26] FMNV include chewing gum, lollipops, jelly beans, and carbonated sodas. Many competitive foods, such as chocolate candy bars, chips, and fruitades (containing little fruit juice), are not considered FMNV, and therefore are allowed to be sold in the school cafeteria during meal times.

Two-thirds of states have weak policies or no policy to address competitive foods. Only twelve states (twenty-four percent) have comprehensive school food and beverage nutrition standards that apply to the whole campus and the whole school day at all grade levels. No states received an A grade, though two states (Kentucky and Oregon) received an A–. Sixteen states received a B grade. Thirteen states received Cs or Ds. Twenty states received Fs.[27]

### Competitive Foods and Their Effect on Children

The Surgeon General's Call to Action to Prevent and Decrease Overweight and Obesity 2001 recommends that "individuals and groups across all settings . . . [adopt] policies specifying that all foods and beverages available at school contribute toward eating patterns that are consistent with the *Dietary Guidelines for Americans*." Unfortunately, the foods available in many schools don't follow this recommendation. For example, approximately twenty percent of schools offer brand-name fast-food items, such as foods from Pizza Hut or Taco Bell, for sale in the cafeteria.[28] Fast foods are typically high in fat, saturated fat, and sodium.

The problems this causes are very apparent. Selling low-nutrition foods in schools contradicts nutrition education and sends children the message that good nutrition is not important.[29] The school environment should reinforce nutrition education in the classroom and model healthy behaviors. Competitive foods also may stigmatize participation in the school meal programs. Children without money to pay for competitive foods may feel self-conscious or embarrassed to receive the reduced-price or free meals available if children with greater financial resources are purchasing competitive foods.

The sale of competitive foods in schools can negatively affect children's diets because the majority of competitive foods are low in nutrients and high in calories, added sugars, and fat.[30] The most common items for sale in vending machines, school stores, and snack bars include soft drinks, sports drinks, fruit drinks that are not one hundred percent juice, salty snacks, candy, and baked goods that are not low in fat.[31]

Both at lunch and during the entire day, students that participate in the NSLP consume significantly more vegetables, milk, and protein-rich foods, and consume less added sugars, soda, and fruit drinks than nonparticipants.[32]

Only two percent of children (two to nineteen years) meet the recommendations for a healthy diet from the Food Guide Pyramid.[33] Only sixteen percent of children meet dietary recommendations for saturated fat.[34] Between 1989 and 1996, children's calorie intake increased by approximately eighty to 230 extra calories per day depending on the child's age and activity level.[35] The increases in calorie intake are driven by increased intakes of foods and beverages high in added sugars.

Teens are snacking more than they used to. The average number of snacks eaten per day increased from 1.6 to 2.0 between 1977 and 1996. Snacks now provide approximately 610 calories to teens' diets each day, versus 460 calories in 1977.[36] Soft-drink consumption by children increased forty percent between 1989 and 1996, from an average of 1.0 to 1.4 servings per day.[37] Fifty-six to eighty-five percent of children drink soda each day (depending on age).[38] Children who drink soft drinks consume more calories (about fifty-five to 190 per day) than kids who do not drink soft drinks.[39] A study conducted by the Harvard School of Public Health found that for each additional can or glass of soda or juice drink a child consumes per day, the child's chance of becoming overweight increases by sixty percent.[40]

Children and adolescents who are obese miss about four school days a month. Frequent absenteeism may lead to lower performance in school.[41]

Consumption of soft drinks can displace low-fat milk and one hundred percent juice from children's diets.[42] Only thirty percent of children consume the recommended number of servings of milk each day, down from forty percent in 1989. Just fourteen percent of children eat the recommended amount of fruit.[43]

Twenty years ago, boys consumed twice as much milk as soft drinks, and girls consumed fifty percent more milk than soft drinks. Today, children consume twice as many soft drinks as glasses of milk.[44] Milk is an important source of calcium to help children build strong bones. Maximum bone mass for women is acquired by age twenty, and low bone density in youth can cause osteoporosis later in life.[45] While a number of different factors cause tooth decay, the American Dental Association concludes that sugared soda increases the risk of dental caries and the low pH of soda can cause tooth erosion.[46]

# HOW YOU CAN IMPROVE
## SCHOOL FOODS AND BEVERAGES

Parents can have an enormous impact on the kinds of foods provided to their kids in schools as long as they get involved. Sometimes a single well-timed effort can serve as a powerful catalyst for change.

In 2002, the Portsmouth, New Hampshire, School Board eliminated drinks high in sugar and caffeine, including soda, sweetened lemonade, iced tea, and fruit drinks, and replaced them with plain and flavored waters and one hundred percent fruit juice. They also eliminated all candy bars and snacks with little nutritional value and replaced them with crackers, Chex mix, raisins, and fruit snacks.

While the school system had been talking about this issue for some time, the catalyst for the change was a letter written to the board by a local pediatrician. In his letter, he criticized the presence of soda and unhealthy snacks in vending machines at schools and pointed out the correlation with rising obesity rates in children.

Here are some of the practical steps you can take to help improve the food choices at the schools in your community:

- Urge your school to adopt strong nutrition standards for foods and beverages sold out of vending machines, school stores, à la carte, at fundraisers, and at other venues outside of the school meal programs. For a model policy, see www.schoolwellnesspolicies.org.
- Work through your school wellness policy committee, PTA, community health council, or other groups to improve school foods and beverages.
- Meet with the person or group that makes decisions on school foods and beverages. For example, school meals are usually the responsibility of school food service, while vending could be managed by the principal, superintendent, food service, or athletic departments.
- Recruit a health professional—e.g., a pediatrician, family doctor, dietitian, nurse, or dentist—to testify before the school board on the connection between childhood obesity and soft drink consumption.
- Encourage your state legislators, governor, school board, superintendent, or other elected official to work to enact legislation to

improve the nutritional quality of foods and beverages sold out of vending machines, à la carte, at fundraisers, and at other school venues.

- Draw your community's attention to the rising problem of child-hood obesity, poor nutrition, and school foods and beverages through media or community events.
- Urge the food service director to promote and serve only low-fat milk at your school.
- Draw attention to junk food in schools by conducting a survey of what is sold in school vending machines and publishing the results.
- Urge your school not to raise money by selling junk food to kids. Volunteer to organize a fundraiser demonstrating healthy ways to raise revenue, such as organizing a car wash, hosting a fun run, or selling raffle tickets. See the website http://www.cspinet.org/nutritionpolicy/fundraiserfactsheet.pdf for a list of healthy, profitable fundraising ideas.
- Work with your school or school district to ensure that only healthy beverages can be sold when the school or district enters into or renews a soft-drink contract.
- Share examples of schools that have changed to healthier options without losing money (see the webpage http://www.cspinet.org/nutritionpolicy/improved_school_foods_without_losing_revenue2.pdf for examples).

Each of these strategies has been used by parents and citizens' groups around the country, often with remarkable success. You can learn more about how to plan and implement these techniques by visiting the CSPI website, where you can find step-by-step advice on how to influence local decision-makers, enlist support from others in your community, and attract attention from the media for your cause.

## WILL SCHOOLS LOSE FUNDS IF COMPETITIVE FOODS REGULATIONS ARE STRENGTHENED?

One of the arguments often raised *against* the implementation of better nutrition standards in schools is an economic one. In times when school budgets—like other government budgets—are often strapped, some legislators and school administrators say that the funds raised through sale of soft drinks and less-than-healthy snacks is an indispensible element in school funding.

According to the USDA and the Centers for Disease Control and Prevention (CDC), "students will buy and consume healthful foods and beverages—and schools can make money from selling healthful options."[47] Their survey of seventeen schools and school districts found that, after improving school foods, twelve schools and districts increased revenue and four reported no change. The food service department of the one school district that initially lost revenue later saw its revenues increase and surpass previous levels.[48]

Also, in two pilot studies that evaluated the financial impact of switching to healthier school foods options, total revenues increased at the majority of schools because meal revenue increases exceeded any losses from the sale of foods and beverages outside of the meals.[49] Here is an example that illustrates the point.

Vista Unified School District in California bought its own vending machines and replaced high-fat and sugar snacks with healthier options, such as yogurt and granola, fruit, and cheese and crackers, and offered less soda and more water, sports drinks, one hundred percent juice, milk, and smoothies. In the first year, the school generated $200,000 more in sales than it had in the previous year.

A middle school and high school in Philadelphia changed their vending machines' beverage contents to remove sports drinks (sodas were already banned) and to include only one hundred percent juice, twenty-five percent juice, and water. Average monthly revenue from the machines did not decrease.

## GOALS TO PURSUE

Once you've decided you're ready to get involved in the effort to improve the foods served at your community's schools, you and your allies will need some specific goals to rally around. Here are a set of goals recommended by CSPI for improving school nutrition programs:

- Set nutrition standards for foods and beverages sold outside of school meals, including out of vending machines, snack bars, à la carte, at fundraisers, in school stores, and at other venues to reduce the availability of low-nutrition foods (soda, chips, candy, etc.) in schools.

- Limit the sale and availability of soft drinks, fruit "drinks" and "ades" (Fruitopia, Snapple, etc.), sports drinks, and other sugary drinks.
- Improve the nutritional quality and "kid-appeal" of school meals.
- Help schools to meet the USDA's nutrition standards for school meals.
- Promote and serve more whole grains and fruits and vegetables.
- Reduce children's intake of saturated fat by making one percent and fat-free milk the standard milks sold in your school.
- Strengthen nutrition education in schools.
- Replace fundraisers that sell candy or other junk food with healthy alternatives.
- Urge your school or school district to give children enough time to eat (the USDA recommends at least twenty minutes for breakfast and thirty minutes for lunch).
- Encourage parents and kids to pack healthy lunches.
- Keep campuses closed during lunch to prevent students from going to fast-food outlets and convenience stores off campus and reduce competition with the school meal programs.
- Reduce junk-food marketing on school campus.
- Keep brand-name fast food out of school cafeterias.
- Implement a policy for offering healthy foods and beverages at school functions, school parties, and staff events.

Choose your own nutrition goals based on the needs and interests of kids and families in your community as well as the current policies in place where you live. Having specific goals will make it easier to mount an effective local campaign on behalf of better nutrition for all families.

## ADVOCATING RESPONSIBLE
## FOOD MARKETING TO CHILDREN

Of course, schools aren't the only place where our children may be eating foods that may be harmful to their long-term health. Every time they visit a mall, stop at a fast-food restaurant on the way to school, pick up a snack before dinner, or take a food break while studying or just hanging out with friends, they have an opportunity to make smart or unhealthy food choices, which over time have a significant impact on their well-being.

It's impossible for parents to act as their kids' "food police," even if they wanted to. The older your kids get, the greater freedom they will have to make their own food choices—and that often means the greater the influence of advertisers, marketers, public relations companies, and other business interests on their diet.

For these reasons, many concerned parents and other citizens who worry about the long-term health of our society—including, for example, the impact on health-care costs—have been advocating greater responsibility on the part of food and media companies when it comes to marketing food to kids.

CSPI is at the forefront of this effort. The following Guidelines for Responsible Food Marketing to Children are for food manufacturers; restaurants; supermarkets; television and radio stations; movie studios; magazines; public relations and advertising agencies; schools; toy and video game manufacturers; organizers of sporting or children's events; and others who manufacture, sell, market, advertise, or otherwise promote food to children. The guidelines provide criteria for marketing food to children in a manner that does not undermine children's diets or harm their health. We hope the guidelines will be helpful to parents, school officials, legislators, community and health organizations, and others who are seeking to improve children's diets.

***Companies should support parents' efforts to foster healthy eating habits in children.***    Parents bear the primary responsibility for feeding their children. However, getting children to eat a healthful diet would be much easier for parents if they did not have to contend with billions of dollars' worth of sophisticated marketing for low-nutrition foods.

Children receive about 6,000 messages from television advertising each year (about half are for food), along with many additional marketing messages from websites, in schools, and in retail stores. Given how often companies communicate with children about food, those who manufacture, sell, and promote food to children have an enormous effect on parents' ability to feed their children a healthful diet.

Parental authority is undermined by wide discrepancies between what parents tell their children is healthful to eat and what marketing promotes as desirable to eat. In addition, while many parents have limited proficiency in nutrition, companies have extensive expertise in persuasive techniques. Companies also have resources to influence children's food choices that parents do not have, such as cartoon characters, contests, celebrities, and toy giveaways.

***Children of all ages should be protected from the marketing of foods that can harm their health.*** The Guidelines for Responsible Food Marketing to Children apply to children of all ages (less than eighteen years of age). Society provides special protections for children, including measures to protect their health, such as requiring use of car safety seats or prohibiting them from buying cigarettes or alcoholic beverages. However, even in the absence of legislative or regulatory requirements, marketers should act responsibly and not urge children to eat foods that could harm their health.

Children are uniquely vulnerable to the marketing of low-nutrition foods. Many children lack the skills and maturity to comprehend the complexities of good nutrition or to appreciate the long-term consequences of their actions. Children of different ages face diverse challenges to healthful eating and different vulnerabilities to food marketing. Young children do not understand the persuasive intent of advertising/marketing and are easily misled. Older children, who still do not have fully developed logical thinking, have considerable spending money and opportunities to make food choices and purchases in the absence of parental guidance.

***Nutrition guidelines.*** Responsible food marketing to children must address not only *how* food is marketed, but also *which* foods are marketed to kids. The Guidelines for Responsible Food Marketing to Children set criteria for which foods are appropriate to market to children. Other guidance regarding marketing to children has focused primarily on marketing techniques. For example, industry's self-regulatory guidelines through the

Children's Advertising Review Unit (CARU) of the Council of Better Business Bureaus address which approaches are appropriate to use in marketing to children. Also, the Federal Trade Commission occasionally takes action against ads deemed unfair or deceptive.

What those approaches fail to address is that most of the food marketed to children is of poor nutritional quality. Changing the way a sales pitch is couched is irrelevant if the product is unhealthy. It hardly matters whether a company markets a candy to children by placing a cartoon character on the package, by promoting it with a contest, or by advertising it on television. What matters is that the marketing encourages children to eat a product of poor nutritional quality that can undermine their diets.

Ideally, companies would market to children only the most healthful foods and beverages, especially those that are typically underconsumed, such as fruits, vegetables, whole grains, and low-fat dairy products.

However, nutrition criteria that would allow only marketing of those foods seem unrealistically restrictive. Instead, we recommend a compromise approach. These guidelines set criteria that allow for the marketing of products that may not be nutritionally ideal but that provide some positive nutritional benefit and that could help children meet the *Dietary Guidelines for Americans* (i.e., that help them to limit their intake of calories, saturated and trans fats, sodium, and refined sugars).

This approach limits the promotion of some foods that are now commonly marketed to children. However, it allows companies to market reasonable alternatives to those products and a wider range of products. It also should provide an incentive for companies to develop and increase demand for foods that are nutritionally better than those that are currently marketed to children.

Some marketing efforts do not promote individual products but instead promote a line of products, one brand within a company, or a whole company. For example, a campaign might encourage children to go to a particular restaurant without marketing a specific menu item. A company logo or spokes-character featured on a hat or website might promote a whole line of products. Companies should not conduct general brand marketing aimed at children for brands under which more than half of the products are of poor nutritional quality, as defined below. If multiple products are shown in an advertisement, if one product does not meet the nutrition criteria below, then the advertisement is considered to promote foods of poor nutritional quality.

**Guidelines for Beverages.**   Low-nutrition beverages should not be marketed to children. These include:

- Soft drinks, sports drinks, and sweetened iced teas
- Fruit-based drinks that contain less than fifty percent juice or that contain added sweeteners
- Drinks containing caffeine (except low-fat and fat-free chocolate milk, which contain trivial amounts of caffeine)

Nutritious/healthful beverages include:

- Beverages that contain at least fifty percent juice and that do not contain added sweeteners
- Low-fat and fat-free milk, including flavored milks and calcium-fortified soy and rice beverages

**Guidelines for Foods.**   Foods marketed to children should meet all of the following criteria (nutritionally poor choices or low-nutrition foods are those that do not meet the criteria):

- Fat: no more than thirty-five percent of total calories, excluding nuts, seeds, and peanut or other nut butters
- Saturated plus trans fats: no more than ten percent of calories
- Added sugars: less than thirty-five percent of added sugars by weight (added sugars exclude naturally occurring sugars from fruit, vegetable, and dairy ingredients)
- Sodium: no more than: 1) 230 mg per serving of chips, crackers, cheeses, baked goods, French fries, and other snack items; 2) 480 mg per serving for cereals, soups, pastas, and meats; 3) 600 mg for pizza, sandwiches, and main dishes; and 4) 770 mg for meals

- Nutrient content: contains one or more of the following: 1) ten percent of the RDI of (naturally occurring/without fortification) vitamins A, C, or E; calcium; magnesium; potassium; iron; or fiber; 2) half a serving of fruit or vegetable; or 3) fifty-one percent or more (by weight) whole grain ingredients

***Portion size limits for foods and beverages.***

- Individual items: no larger than the standard serving size used for Nutrition Facts labels (except for fruits and vegetables, which are exempt from portion size limits)
- Meals: no more than one-third of the daily calorie requirement for the average child in the age range targeted by the marketing

***Marketing techniques.*** When marketing foods to children, companies should:

- Support parents' efforts to serve as the gatekeepers of sound nutrition for their children and not undermine parental authority. Marketers should not encourage children to nag their parents to buy low-nutrition foods
- Depict and package/serve food in reasonable portion sizes and not encourage overeating directly or indirectly
- Develop new products that help children eat healthfully, especially with regard to nutrient density, energy density, and portion size
- Reformulate products to improve their nutritional quality, including adding more fruits, vegetables, and whole grains, and reducing portion sizes, calories, sodium, refined sugars, and saturated and trans fats
- Expand efforts to promote healthy eating habits consistent with the *Dietary Guidelines for Americans* and to promote healthful products, such as fruits, vegetables, whole grains, and low-fat milk. Do not portray healthful foods negatively
- Do not advertise nutritionally poor choices during television shows: 1) with more than fifteen percent of the audience under age twelve; 2) for which children are identified as the target audience by the television

station, entertainment company, or movie studio; or 3) that are kid-oriented cartoons

- Do not use product or brand placements for low-nutrition foods in media aimed at children, such as movies, television shows, video games, websites, books, and textbooks
- Only offer premiums and incentives (such as toys, trading cards, apparel, club memberships, products for points, contests, reduced-price specials, or coupons) with foods, meals, and brands that meet the nutrition criteria described above
- Use/allow licensing agreements or cross-promotions (such as with movies, television programs, or video games) or use cartoon/fictional characters or celebrities from television, movies, music, or sports to market to children only those foods that meet the above nutrition criteria. This includes depictions on food packages, in ads, as premiums, and for in-store promotions
- Do not put logos, brand names, spokes-characters, product names, or other marketing for low-nutrition foods/brands on baby bottles, children's apparel, books, toys, dishware, or other merchandise made specifically for children
- Incorporate into games (such as board, Internet, or video games), toys, or books only those products and brands that meet the nutrition criteria
- Use sponsorship of sporting, school, and other events for children only with brands and foods that meet the above nutrition criteria
- Do not exploit children's natural tendency to play by building entertainment value into low-nutrition foods (for example, products such as mechanical lollipops, food shaped like cartoon characters, or sugary drink mixes that turn to surprise colors)

***Additional guidance for schools.***    Schools are a unique setting. Parents entrust their children into schools' care for a large proportion of children's waking hours. Also, schools are dedicated to children's education and are supported by tax dollars. Companies should support healthy eating in schools and not market, sell, or give away low-nutrition foods or brands anywhere on school campuses, including through:

- Logos, brand names, spokes-characters, product names, or other product marketing on/in vending machines; books, curricula, and other educational materials; school supplies; posters; textbook covers; and school property such as scoreboards, signs, athletic fields, buses, and buildings
- Educational incentive programs that provide food as a reward (for example, earning a coupon for a free pizza after reading a certain number of books)
- Incentive programs that provide schools with money or school supplies when families buy a company's food products
- In-school television, such as Channel One
- Direct sale of low-nutrition foods
- Free samples or coupons
- School fundraising activities
- Banner ads or wallpaper on school computers

**Additional guidance for retail stores (such as grocery, toy, convenience, and video stores).**

- Replace low-nutrition foods with more healthful foods or nonfood items at checkout aisles or counters
- Do not position in-store displays for low-nutrition foods or place low-nutrition products on shelves at young children's eye level
- Cluster cookies, chips, candy, soda, and other food categories that are predominantly of poor nutritional quality in a few designated aisles of grocery stores to allow parents to skip those aisles if they choose

**Companies should not use the following approaches to market any foods (irrespective of the nutritional quality of the food being marketed).**

- Should not show children engaged in other activities (like skateboarding, playing soccer, watching television, etc.) while eating
- Should not mislead children regarding the emotional, social, or health benefits of a product or exploit children's developmental vulnerabilities and emotions to market any food

- Should not use physical activity or images of healthful foods (such as fruits and vegetables) to market any low-nutrition food
- Should not link children's self-image to the consumption of any foods/brands, use peer pressure, or arouse unrealistic expectations related to consuming/purchasing a food (for example, implying that a child will be more physically fit, more accepted by peers, happier, or more popular if he buys a food/brand or goes to a certain restaurant)
- Should not market any food by modeling rebellion against parents or by portraying parents, teachers, or other authority figures in negative roles
- Should not suggest that an adult who buys a child a certain product is more loving, generous, or otherwise better than an adult who does not

Experience shows that American companies are very conscious about their image as good corporate citizens. They respond to pressure from consumers because they want to be seen as "doing the right thing." You, your friends, and your neighbors can play an important role in changing the behavior of food marketing companies just by making your opinions known. Write letters to company executives; sound off at community meetings and in the letters columns of newspapers and magazines; convey your feelings to local retailers and to media outlets that carry food advertising aimed at young people.

Above all, vote with your spending. Make a point of tracking which companies behave responsibly when it comes to marketing foods to kids and buy the products they offer. Avoid the products sold by companies that you see as being "repeat offenders" in peddling unhealthy foods to young people.

## CALLING FOR NUTRITION LABELING AT FAST-FOOD AND OTHER CHAIN RESTAURANTS

Yet another important influence on children's diets is restaurant foods. Studies show that Americans are eating an ever-increasing share of their meals away from home. That means it's more vital than ever that restaurants let us know what we are eating when we visit a fast-food restaurant or other food outlet.

Menu labeling is one important way to allow families to make informed choices when eating out. In recent years, more and more states and localities are adopting ordinances requiring that restaurants provide nutrition information about the foods they serve. You can help by urging the adoption of a menu-labeling policy in the state or city where you live. Here are some of the important facts that help to explain why menu labeling is necessary and how it will benefit us.

American adults and children consume on average one-third of their calories from eating out. The average American eats out four meals a week; that is enough to lead to overconsuming calories not just on the day the person eats out, but also to exceed calorie requirements over the course of a whole week.

Studies link eating out with obesity and higher calorie intakes. Children eat almost twice as many calories when they eat a meal at a restaurant compared to a meal at home. When eating out, people eat more saturated fat and fewer nutrients, such as calcium and fiber, than at home. For example, one order of cheese fries with ranch dressing contains 3,010 calories, a large movie theater popcorn with "butter" topping has more than 1,600 calories, and a café mocha and pastry from Starbucks provide more than 1,000 calories.

Since 1994, the Nutrition Labeling and Education Act (NLEA) has required food manufacturers to provide nutrition information on nearly all packaged foods. Three-quarters of adults report using nutrition labels on packaged food, and using labels is associated with eating more healthful diets. However, NLEA explicitly exempts restaurants. Studies show that providing nutrition information at restaurants can help people make lower-calorie choices. Menu labeling allows Americans to exercise personal responsibility and make informed choices for the growing part of their diets that is consumed away from home.

The logic behind menu labeling is compelling. A basic tenet of the free market is that consumers have a right to information. Companies are required to provide information on the fuel-efficiency of cars, what clothes are made of, care instructions for clothing, and energy and water consumption of certain home appliances.

A labeling requirement at chain restaurants is even more compelling. People need nutrition information to manage their weight and reduce the risk of or manage heart disease, diabetes, or high blood pressure, which are leading causes of death, disability, and high health-care costs. What's more, people *want* nutrition information from restaurants; seventy-eight percent of Americans support menu labeling.

Despite these compelling facts, menu labeling has not been adopted voluntarily by most of America's popular restaurant chains. Half of large chain restaurants do not provide any nutrition information to their customers. And restaurants that do provide nutrition information generally do so in ways that are not visible to customers when ordering. For example, they use websites, which require computers and Internet access before going out to eat; tray liners or fast-food packages, which are not visible to customers until after they order; or brochures and posters, which are easy to overlook or may not be available at all. As a result, one study found that less than four percent of customers saw the nutrition information in chain restaurants that provided it. Nutrition information must be *on the menu* to be useful.

In addition to making it possible for consumers to make smarter choices, menu labeling is likely to spur nutritional improvements in restaurant foods. A key benefit of mandatory nutrition labeling on packaged foods has been the reformulation of existing products and the introduction of new nutritionally improved products. For example, trans fat labeling on packaged food led many companies to reformulate their products to remove trans fat. Similar improvements in restaurant foods will probably happen when menu labeling becomes the norm.

The National Academies' Institute of Medicine recommends that restaurant chains "provide calorie content and other key nutrition information on menus and packaging that is prominently visible at point of choice and use" (2006). The Food and Drug Administration, Surgeon General, U.S. Department of Health and Human Services, National Cancer Institute, and American Medical Association also recommend providing nutrition information at restaurants.

When menu labeling is proposed in your community, support the effort. Let your local legislators and municipal, county, and state executives know that you favor menu labeling rules. You can show your support by writing a letter or email, making a phone call, or meeting with key policymakers or writing a letter to the editor of your local paper.

## SERVING YOUR KIDS HEALTHY SNACKS

Of course, there are plenty of things individual families can do *immedi-ately* to improve the way they eat, in addition to working for stronger policies by schools, food marketers, and restaurant chains. In this sec-tion, we'll focus on a particular area of concern to millions of parents—how to choose snacks that enhance rather than undermine kids' health.

Snacks play a major and growing role in children's diets. Between 1977 and 1996, the number of calories that children consumed from snacks increased by 120 calories per day. That's why serving healthy snacks to children is important to providing good nutrition; supporting lifelong healthy eating habits; and helping to prevent costly and poten-tially disabling diseases, such as heart disease, cancer, diabetes, high blood pressure, and obesity.

Here are some ideas for parents, caregivers, teachers, and program di-rectors for serving healthy snacks and beverages to children in the class-room, in after-school programs, at soccer games, and elsewhere. Some ideas may be practical for large groups of children, while other ideas may only work for small groups, depending on the work and cost involved.

### *Fruits and Vegetables*

Most of the snacks served to children should be fruits and vegetables be-cause most kids do not eat the recommended five to thirteen servings of fruits and vegetables each day. Eating fruits and vegetables lowers the risk of heart disease, cancer, and high blood pressure. Fruits and vegeta-bles also contain important nutrients like vitamins A and C and fiber.

Serving fresh fruits and vegetables can seem challenging. However, good planning and the growing number of shelf-stable fruits and veg-etable products on the market make it easier. Though some think fruits and vegetables are costly snacks, they are actually less costly than many other less-healthful snacks on a per-serving basis. According to the U.S. Department of Agriculture, the average cost of a serving of fruit or veg-etable (all types—fresh, frozen, and canned) is $ .25 per serving. This is a good deal compared with a $ .69 single-serve bag of potato chips or an

$ .80 candy bar. Try lots of different fruits and vegetables and prepare them in various ways to find out what your kids like best.

**Fruit.** Fruit is naturally sweet, so most kids love it. Fruit can be served whole, sliced, cut in half, cubed, or in wedges. Canned, frozen, and dried fruits often need little preparation. Your options include:

apples (it can be helpful to use an apple corer)
apricots
bananas
blackberries
cantaloupe
cherries
grapefruit
grapes (red, green, or purple)
honeydew melon
kiwis (cut in half and give each child a spoon to eat it)
mandarin oranges
mangoes
nectarines
oranges
peaches
pears
pineapples
plums
raspberries
strawberries
tangerines
watermelon

Also consider these healthful ways of serving fruit:

- Applesauce, fruit cups, and canned fruit: these have a long shelf life and are low-cost, easy, and healthy if canned in juice or light syrup. Examples of unsweetened applesauce include Mott's Natural Style and Mott's Healthy Harvest line. Dole and Del Monte offer a variety of single-serve fruit bowls.

- Dried fruit: try raisins, apricots, apples, cranberries, pineapple, papaya, and others with little or no added sugars.
- Frozen fruit: try freezing grapes or buy frozen blueberries, strawberries, peaches, mangoes, and melon.
- Fruit salad: get kids to help make a fruit salad. Use a variety of colored fruits to add to the appeal.
- Smoothies: blend fruit with juice, yogurt or milk, and ice. Many storebought smoothies have added sugars and are not healthy choices.
- Deliveries: deliveries of fresh fruit or platters of cut-up fruit are a convenient option offered by some local grocery stores.

Fruit products to approach with caution include:

- Fruit snacks: some brands of fruit snacks are more like candy than fruit and should be avoided due to their high content of added sugars and lack of fruit. Brands to *avoid* include Fruit Roll-Ups, Farley's Fruit Snacks, Sunkist Fruit Gems, Starburst Fruit Chews, Mamba Fruit Chews, Jolly Rancher Fruit Chews, Original Fruit Skittles, and Amazin' Fruit Gummy Bears. Try Natural Value Fruit Leathers and Stretch Island Fruit Leathers, which come in a variety of flavors and don't have added sugars.
- Popsicles: most so-called "fruit" popsicles have added sugars and should be reserved for an occasional treat. Look for popsicles made from one hundred percent fruit juice with no added caloric sweeteners, such as Breyers or Dole "No Sugar Added" fruit bars.

**Vegetables.**    Vegetables can be served raw with dip or salad dressing. Look for the following possibilities in season:

broccoli
carrot sticks or baby carrots
cauliflower
celery sticks

cucumber

peppers (green, red, or yellow)

snap peas

snow peas

string beans

tomato slices or grape or cherry tomatoes

yellow summer squash slices

zucchini slices

To make veggies extra appealing, consider these options:

- Dips: try low-fat salad dressings, like fat-free ranch or thousand island, store-bought light dips, bean dips, guacamole, hummus (which comes in dozens of flavors), salsa, or peanut butter.
- Salad: make a salad or set out veggies like a salad bar and let the kids build their own salads.
- Soy: edamame (pronounced "eh-dah-MAH-may") are fun to eat and easy to serve. (Heat frozen edamame in the microwave for about two to three minutes).
- Veggie pockets: cut whole-wheat pitas in half and let kids add veggies with dressing or hummus.
- Ants on a log: let kids spread peanut butter on celery (with a plastic knife) and add raisins.

### Healthy Grains (Bread, Crackers, Cereals, Etc.)

Though most kids eat plenty of grain products, too many of those grains are cookies, snack cakes, sugary cereals, Rice Krispies treats, and other refined grains that are high in sugars or fat. Try to serve mostly whole grains, which provide more fiber, vitamins, and minerals than refined grains. In addition, try to keep the sugars to less than thirty-five percent by weight and the saturated plus trans fats low (i.e., less than ten percent of calories, or about one gram or less per serving).

Note: cookies, snack cakes, and chips should be saved for occasional treats, given their poor nutritional quality.

Some grain-based treats to serve kids at snack time include:

- Whole wheat English muffins, pita, or tortillas: stuff pitas with veggies or dip them in hummus or bean dip.
- Breakfast cereal: Either dry or with low-fat milk, whole grain cereals such as Cheerios, Grape-Nuts, Raisin Bran, Frosted Mini-Wheats, and Wheaties make good snacks. Look for cereals with no more than thirty-five grams of sugars by weight (or no more than ten grams of sugars per serving).
- Crackers: whole-grain crackers such as Triscuits, which come in different flavors, or thin crisps (or similar woven wheat crackers), Kalvi Rye crackers, or whole-wheat Matzos can be served alone or with toppings, like low-fat cheese, peanut butter, or low-fat, reduced-sodium luncheon meat.
- Rice cakes: look for rice cakes made from brown (whole grain) rice. They come in many flavors and can be served with or without toppings.
- Popcorn: look for low-fat popcorn in a bag or microwave popcorn. Or you can air pop the popcorn and season it, e.g., by spraying it with vegetable oil spray and adding parmesan cheese, garlic powder, or other non-salt spices.
- Baked tortilla chips: baked tortilla chips are usually low in fat and taste great with salsa and/or bean dip. Look for brands with less sodium.
- Granola and cereal bars: look for whole-grain granola bars that are low in fat and sugars, such as Barbara's Granola Bars (cinnamon raisin, oats and honey, and carob chip flavors), Nature Valley Crunchy Granola Bars (cinnamon, oats 'n' honey, maple brown sugar, and peanut butter flavors), Nature Valley Chewy Trail Mix Bars (fruit and nut flavor), and Quaker Chewy Granola Bar (peanut butter and chocolate chunk flavor).
- Pretzels, breadsticks, and flatbreads: these low-fat items can be offered as snacks now and then. However, most of these snacks are not whole grain, and most pretzels are high in salt.

### Low-Fat Dairy Foods

Dairy foods are a great source of calcium, which can help to build strong bones. However, dairy products also are the biggest sources of artery-clogging saturated fat in kids' diets. To protect children's bones and hearts, make sure all dairy foods served are low-fat or fat-free.

- Yogurt: look for brands that are low-fat or fat-free, moderate in sugars (no more than about thirty grams of sugars in a six-ounce cup), and high in calcium (at least twenty-five percent of daily value [DV] for calcium in a six-ounce cup). Examples include Danimals Drinkable Low-Fat Yogurt; Go-Gurt by Yoplait; or cups of low-fat or non-fat yogurt from Stonyfield, Dannon, Horizon, and similar store brands. Low-fat or non-fat yogurt also can be served with fresh or frozen fruit or low-fat granola.
- Low-Fat cheese: cheese provides calcium, but often its saturated fat price tag is too high. Cheese is the number-two source of heart-damaging saturated fat in children's diets. Even with low-fat and reduced-fat cheese, be sure to serve with other foods like fruit, vegetables, or whole grain crackers. Choose reduced-fat cheeses such as Trader Joe's Armenian Style Braided; Borden or Sargento Light Mozzarella string cheese; Frigo Light Cheese Heads; Kraft Twist-Ums; Polly-O Twisterellas; the Laughing Cow's Light Original Mini Babybel; or Cabot fifty percent Light Vermont Cheddar.
- Low-Fat pudding and frozen yogurt: low-fat or fat-free pudding and frozen yogurt should be served only as occasional treats because they are high in added sugars.

### Other Snack Ideas

- Nuts: because nuts are high in calories, it is best to serve them along with another snack such as fruit. A small handful of nuts is a reasonable serving size. Examples include peanuts, pistachios, almonds, walnuts, cashews, or soy nuts. Look for nuts that are unsalted. WARNING: A small but growing number of kids have severe peanut and/or tree nut allergies. Before bringing in peanuts, peanut butter, or other nuts as a snack, check to make sure none of the children has an allergy.
- Trail mix: trail mixes are easy to make and store well in a sealed container. Items to include: low-fat granola, whole-grain cereals,

peanuts, cashews, almonds, sunflower seeds, pumpkin seeds, and dried fruits like raisins, apricots, apples, pineapple, or cranberries.

- Luncheon meat: choose lower-fat, reduced-sodium brands of turkey, ham, and roast beef and serve with whole-wheat bread, pita, tortillas (as a wrap sandwich), or crackers. Cut sandwiches in half to make snack-sized portions.

### Healthy Beverages

- Water: water should be the main drink served to kids at snack times. Water satisfies thirst and does not have sugar or calories. (Plus, it is low-cost for caregivers!) If kids are used to getting sweetened beverages at snack times, it may take a little time for them to get used to drinking water.
- Seltzer: carbonated drinks such as seltzer, sparkling water, and club soda are healthy options. They do not contain the sugars, calories, and caffeine of sodas. Serve them alone or try making "healthy sodas" by mixing them with equal amounts of one hundred percent fruit juice.
- Low-fat and fat-free milk: milk provides key nutrients, such as calcium and vitamin D. Choose fat-free (skim) or low-fat (one percent) milk to avoid the heart-damaging saturated fat found in whole and two percent (reduced-fat) milk. It is best to serve fat-free versions of chocolate, strawberry, or other flavored milks to help balance the extra calories coming from added sugars. Single-serve containers of chocolate or other flavored whole or two percent milk drinks can be too high in calories (400–550 calories) and saturated fat (one-third of a day's worth) to be a healthy beverage for kids.
- Soy and rice drinks: for children who prefer not to drink cow's milk, calcium-fortified soy and rice drinks are good choices.
- Fruit juice: try to buy one hundred percent fruit juice and avoid the added sugars of juice drinks, punches, fruit cocktail drinks, or lemonade. Drinks that contain at least fifty percent juice and no additional caloric sweeteners are also healthful options. To find one hundred percent juice, look at beverage nutrition labels for the percentage of the beverage that is juice. Orange, grapefruit, and

pineapple juices are more nutrient-dense and are healthier than apple, grape, and pear juices. Many beverages such as Capri Sun; V8-Splash; Tropicana Twisters; Sunny Delight; Kool-Aid Jammers; Hi-C; or juice drinks from Very Fine, Welch's, or Snapple are easily mistaken for juice. However, those beverages are more like soda than juice—they are merely sugar water with a few tablespoons of added juice.

Fruit juice can be rich in vitamins, minerals, and cancer-fighting compounds. However, it is high in calories. The American Academy of Pediatrics recommends that children ages one to six years old drink no more than six ounces (one serving) of juice a day and children ages seven to eighteen years old drink no more than twelve ounces (two servings) of juice a day.

A note about sugary soft drinks (soda, sweetened tea, lemonade, and juice drinks): children who drink more sweetened drinks consume more calories and are more likely to be overweight than kids who drink fewer soft drinks. Soft drinks also displace healthful foods in kids' diets like milk, which can help prevent osteoporosis. In addition, soda pop can cause dental cavities and tooth decay.

Serving healthy snacks to your kids and to the kids whose lives you touch can be a great way not just to improve their nutrition in the short term, but also to help them develop good eating habits that will benefit them throughout their lives.

# Another Take | CHILDHOOD OBESITY
## THE CHALLENGE

### By the Robert Wood Johnson Foundation

The Robert Wood Johnson Foundation is the nation's largest philanthropy devoted exclusively to improving the health and health care of all Americans. In addressing critical issues facing this country, it seeks to advance individual health as well as better every person's health care—how it's delivered, how it's paid for, and how well it does for patients and their families. Among the social changes that the foundation has spearheaded in its history are the establishment of 911 emergency telephone service throughout the United States, widespread public awareness of the health consequences of tobacco use and significant reduction in smoking rates, and recognition and support of the field of end-of-life health care. Today, the foundation is focusing much of its efforts on reversing the childhood obesity epidemic. For more information, visit the foundation's website, located at http://www.rwjf.org/childhoodobesity/index.jsp.

Childhood obesity is a serious public health problem in the United States. Over the past three decades, obesity rates have soared among all age groups, increasing more than four times among children ages six to eleven. Today, more than twenty-three million children and teenagers are overweight or obese. That's nearly one in three young people. Even among ages two to five, a quarter of children are now overweight or obese. Among certain racial and ethnic groups, the rates are still higher.

The ramifications are alarming. If we don't succeed in reversing this epidemic, we are in danger of raising the first generation of American children who will live sicker and die younger than their parents' generation.

Preventing obesity during childhood is critical because habits formed during youth frequently continue well into adulthood:

- Research shows that obese adolescents have up to an eighty percent chance of becoming obese adults. Overweight and obese children are at higher risk for a host of serious illnesses, including heart disease, stroke, asthma, and certain types of cancer.
- Increasing numbers of children are being diagnosed with health problems once considered to be adult ailments, including high blood pressure, type 2 diabetes, and gallstones.
- Obesity poses a tremendous financial threat to our economy and our health-care system. It's estimated that the obesity epidemic costs our nation $117 billion annually in direct medical expenses and indirect costs, including lost productivity. Childhood obesity alone carries a huge price tag—up to $14 billion annually in direct health-care costs.

How did we get to this point? There's a simple explanation for the childhood obesity epidemic: our children are consuming far more calories than they burn. Today's obese teenagers consume between 700 and 1,000 more calories per day than what's needed for the growth, physical activity, and body function of a normal-weight teen. Over the course of ten years, that "energy gap" is enough to pack an average of fifty-eight extra pounds on an obese adolescent.

As a society, we've dramatically altered the way we live, eat, work, and play—creating an environment that fuels obesity:

- On average, today's young people spend more than four hours per day using electronic media, including television, DVDs, and video games.
- A generation ago, about half of all school-aged children walked or biked to school. Today, nearly nine out of ten are driven to school. And once they get there, there aren't many opportunities for exercise—fewer than four percent of elementary schools provide daily physical education.
- At the same time, children are eating more unhealthy foods in ever-larger sizes. In recent decades, the typical calorie content of menu

items like French fries and sodas has increased approximately fifty percent. Children consume these high-calorie, low-nutrient foods not only in restaurants, but also in their homes and schools.

- In communities hardest hit by obesity, families frequently have little access to affordable healthy foods. There often are no grocery stores, only convenience marts that rarely stock fresh fruits and vegetables. There aren't enough safe places for kids to play or programs to help them be physically active every day.

To reverse the childhood obesity epidemic, we must remove these barriers by creating policies and environments that provide families with better access to healthy foods and opportunities for physical activity.

## THIRTEEN

# PRODUCE TO THE PEOPLE
## *A PRESCRIPTION FOR HEALTH*

*By Preston Maring*

---

Dr. Preston Maring is the associate physician-in-chief at the Kaiser Permanente East Bay Medical Center in Oakland, Kaiser Permanente's first hospital location. He is responsible for tertiary care services planning and development for Oakland and Richmond's 270,000 health plan members as well as members from around the Northern California region. He is board certified in obstetrics and gynecology and received his MD from the University of Michigan in 1971. He completed his residency in obstetrics and gynecology at the Kaiser Permanente Medical Center in Oakland in 1974.

Dr. Maring is an enthusiastic cook who was supported by the Kaiser Permanente leadership at the Oakland Medical Center to start a farmers' market for the benefit of staff, members, visitors, the community around the medical center, and the farmers. The success of this first market has inspired Kaiser Permanente employees at multiple facilities in four states to create their own markets. The innovation and spirit of the local facility sponsors has resulted in different market models, community outreach, and a program-wide focus on healthy eating. Dr. Maring has personally supported the development of many of these markets. More recently, he has worked with Kaiser Permanente and the Community Alliance with Family Farmers to help create a system that sources produce for inpatient meals from small family farmers with a focus on increasing the utilization of sustainably produced foods.

---

W alking through the lobby of my hospital in the summer of 2002, I noticed a vendor selling jewelry and other accessories. Jewelry for sale in a hospital didn't fit exactly with the genetic makeup of our medical care program at Kaiser Permanente, but the commercial activity made me think about another type of market, one that would fit our focus on preventing disease and then treating illnesses the best way possible the first time they do occur.

What about a farmers' market? The location, where thousands of people gather each day, made perfect sense. The farmers' markets in my community at the time were like those found around the country—set up on a closed-off street or a public courtyard. As far as I knew, there weren't any hospital-based farmers' markets in California and very few across the country.

I had worked much of my life as a primary-care women's health physician and knew nothing about starting a farmers' market. As I had always done in my medical practice when I wasn't sure about something, I sought out the advice of someone who actually knew the answers.

I found that person on a sunny Sunday afternoon at Oakland's Jack London Square farmers' market while shopping with my wife and niece. One of the farmer's stalls in particular caught my eye that day because its proprietor was selling vegetables that could be used in all kinds of cuisine. He had Chinese long beans, pasilla peppers, and Belgian endive. The shoppers at his stand represented many different cultures and looked just like my patients or my hospital's workforce, in a city whose population is one of the most diverse in the country.

I stopped to talk with the farmer, a little hesitant because I was going to ask him a question he may not have been asked before. But his answer started me on a journey. He said with a smile that the man I needed to talk to was standing just ten feet away.

That man was John Silveira, director of the nonprofit Pacific Coast Farmers Market Association, which managed twenty-two community markets around the Bay Area at the time. I introduced myself and asked him if he would consider starting a farmers' market at my hospital. He thought just for an instant, then said, "Your mission and my mission are the same. We both work to help the health of the community."

On a breezy, sunny day the following May, our hospital's farmers' market opened for business. It was like a block party out in front of the medical center. Staff, patients, visitors to the hospital, and neighbors from the surrounding community celebrated with huge, luscious organic strawberries. The Lone Oak Ranch from Reedley, California, a fourth-generation, first-generation organic farm, offered early season stone fruits; Happy Boys Farms displayed an amazing array of spring vegetables. I had never seen such beautiful spring onions.

To this day, I still pinch myself when I come to work on Fridays and see the tents and tables being set up. Marlene Gonzalez of Lone Oak radiates joy when you greet her as the market gets underway. She loves to do what she does, and it's felt by all of us who have the privilege to shop at her stand.

## YOU ARE WHAT YOU EAT

Over my more than thirty-seven years practicing medicine at the Kaiser Permanente Medical Center in Oakland, California, it has become increasingly clear that what people eat is the major determinant of their personal health. Of course it's critical to never smoke again or not start in the first place. The recommended thirty minutes per day of moderate exercise is important. People who move more their whole lives are much more likely to be able to move more when they are older. Bike helmets and seatbelts are a must. The dental hygienist will only speak approvingly if you floss daily. But when about thirty-five percent by weight of all the food eaten in the United States is either fast food or pizza, it makes sense for health-care providers and medical-care programs to focus attention where it's possible to make a big difference.

Incentives at Kaiser Permanente are similar in some ways to those of a New York City cab driver. A taxi cab ride from the John F. Kennedy airport to Manhattan costs a fixed rate of $45 plus tip. The incentives are perfectly aligned for the cabbie and the passenger—find the shortest route to the destination that will get the passenger there safely in the shortest amount of time. Same thing at Kaiser Permanente's medical care program. One of the most direct routes to good health and to prevent disease in the first place has healthy foods offered along the road. Having a farmers' market where people come to receive health care and preventive services has turned out to be one of the very best ways to focus attention on the role of good food in good health.

## FARMERS' MARKETS SPREAD FAR AND WIDE

Word spread fast throughout the regions of Kaiser Permanente in nine states and the District of Columbia. Before long there were farmers' markets springing up at facilities near and far away from Oakland. Ashlyn Izumo, one of the staff at Honolulu Kaiser Permanente, emailed me just a few months after my market started that she had found seven farmers willing to establish a market at her hospital. The Baldwin Park facility in Southern California partnered with a local agricultural college to sell all the produce that was grown at the school. The truck had to go back for more on opening day. Today there are thirty-two markets in four of our regions.

The markets became a focal point for change, even though support for them is not part of anyone's job description. People saw opportunities and then tried to make them real. Terri Simpson-Tucker and colleagues at the San Jose Medical Center partnered with the farmers' market association to outfit a mobile cooking demonstration van that has done about a hundred community events each year. Lucila Santos, a food-services leader in Kaiser Permanente's Southern California region, started cooking classes for children and their parents and for teenagers by themselves. Ruth Conley, of the Kaiser Permanente Watts Counseling and Learning center, helped start a Kaiser Permanente–sponsored market in a community park in her community.

## MARKETS BREED COMMUNITY INVOLVEMENT

Our farmers' markets have helped us to reach out into our local communities. For example, the Oakland hospital farmers' market has hosted field trips from elementary school children and juniors from Oakland High School environmental sciences and physiology classes. Because of the diversity of the students, the conversations at the market during a high school visit in the spring of 2008 covered many different cuisines. We discussed an option for potatoes other than mashing or frying them. I was amazed that these young people were willing to listen to me talk about roasting red potatoes with fennel and onions.

Colleagues and I then fed them lunch in our hospital's garden area while we talked about organic foods and the challenges of growing them. They also had a chance to talk with various health professionals about their work. I asked my colleagues to simply share what they found exciting every day about their jobs and what kind of training was needed post–high school for their particular profession. I believe I will see one young man someday back at Kaiser Oakland as an RN in a critical-care area.

For me, the best part of the day was having a tall young man come back for his third big helping of Caesar salad, look at me, and say, "I don't eat salad." One step at a time.

## FARMERS' MARKETS AND HEALTH

People have asked if I thought the hospital-based markets were actually helping make improvements in people's health. There's no easy way to measure the impact, but there are powerful stories. After our market first opened, I started writing prescriptions for an arugula salad with a Meyer lemon vinaigrette (one part lemon juice to one or two parts extra virgin olive oil and a little salt) for my patients. One of my patients started shopping at the market every Friday after a long life of eating fast foods and saw real changes in her weight and health. One of our hospital engineers credits his now loose-fitting work clothes to the market. It's hard to describe how much fun it is to have him come up to me in the basement to tell me about a raspberry vinaigrette he had made for a big salad.

In 2005, we surveyed about 1,200 people who shopped at seventeen of our markets. Seventy-one percent said they were eating more fruits and vegetables because of the convenient markets at the hospitals, and sixty-three percent said they were trying new fruits and vegetables, even little-known varieties such as kohlrabi, which I had never tried before myself. With only about ten percent of Americans eating five or more servings of fruits and vegetables daily, each farmers' market has a big potential customer base. If all of us met the fruit and vegetable consumption recommendations, I've heard it would take two million more acres to grow them. It sounds like changing America's eating habits could create many jobs and protect a lot of land from development.

## FROM FARM TO HOSPITAL TRAY

Focus on fresh, local produce led me to think about the food served to our patients on hospital trays. In 2006, there were nineteen Northern California Kaiser Permanente hospitals serving about 6,000 meals per day to hospitalized patients. We studied the origins of about sixty fruits and vegetables used in our meals and found that about forty percent were coming from outside of California. Our patients like pineapple and they like mixed leafy greens in their salads in the middle of winter—not just parsnips and rutabaga. We are still going to ship in a lot of produce. But we found that we were purchasing red seedless grapes from halfway around the world when they grow in Fresno, and ordering asparagus in October when it grows here in California as soon as the ground temperature reaches sixty-five degrees. What we didn't know was how to connect a big institutional purchaser like Kaiser Permanente to the local agricultural system.

We found the advice we were looking for from the Community Alliance with Family Farmers (www.CAFF.org), California's largest and oldest nonprofit committed to the support of the small farm economy. Could they help link us up with family farmers to supply at least some of the foods for our patients' trays? After six months of research and planning, we started on a new path together.

Beginning in August 2006, Kalu Afu, a twenty-nine-year employee at Food Service Partners, Kaiser Permanente's inpatient food commissary, was able to place cherry tomatoes grown by Choua Vang, a Hmong farmer with nine leased acres in Fresno, on the dinner salads for patients in our nineteen hospitals. Little did I know at the time that this was possible because of work being done on the other side of the country.

After our hospital food project had already started, I had the privilege of attending a USDA Farmers' Market Summit in Washington, D.C., where I met people from inside and outside of government who were working to support the existing and new small farmers. There are clearly several degrees of separation from policy in Washington to the frontline: The Office of Refugee Settlement in D.C., which helps Hmong immigrants—many of whom are excellent farmers—provided funding to the Fresno Office of Economic Opportunity. Another D.C. organization, the Institute for Social and Economic Development, awarded a grant to the Growers Collaborative, the part of the Community Alliance with Family Farmers that connected us to Mr. Vang's tomatoes.

This small distribution system has been bringing us sixty tons of fruits and vegetables from local family farmers each year. And with the ongoing backbone of support from Kaiser Permanente's institutional purchasing, the Growers Collaborative has also been able to expand its customer base and now delivers to twenty-four hospitals, thirty Bon Appétit sites, local universities, elementary schools, high schools, and social service organizations. Their work initially helped support twelve farmers; they are now working with ninety-seven.

One such farmer who has benefited from the Growers Collaborative's work is Roberto Rodriquez, a strawberry farmer in Watsonville, California, with thirty-five acres of land. Six years ago he had a daughter and started growing some of his acreage organically so she wouldn't be exposed to pesticides. The luscious berries from the fifteen acres that are now organic can be purchased at our hospital's farmers' market between March and October. In addition, he has a standing order of 130 dozen pints per week for our hospitalized patients' desserts. Because of this business, he reports that he's hired five new farmworkers.

Observing this gratifying progress, my colleagues and I were looking for additional ways to get good food to people in our system. Our farmers' markets only serve the larger medical centers that have a few thousand staff and patients on-site every day. What about all those other smaller medical office satellites, office buildings, and other regional buildings housing the pharmacy's central supply, the regional laboratory, and optical lab? What about the people at the facilities with markets who are too busy to shop at the market or are in the operating room all day?

A few of our facilities have their own community-supported agriculture (CSA) deliveries coming in from local farmers that have sufficient acreage and the biodiversity of crops needed. Because there are many farmers not meeting those criteria, we have looked for ways to offer "bundled" CSAs or "collaborative" CSAs. The new "Best of the Market" program at some of the farmers' market facilities provides $10 and $20 bags of what's best each week to be delivered up into the hospitals. Another model partners multiple family farmers with an existing fruit delivery service to get their vegetables and fruit delivered into office buildings by subscription. Some Kaiser Permanente employees really like getting a commuter-friendly box of local produce delivered to their desk. We hope that increased focus on healthy foods through employee wellness programs will help encourage more people to make the change.

Whether it's one patient, one nurse, one doctor, or one farmer at a time, it's important to continue the effort to expand the access to fresh, locally grown food. My experience at Kaiser Permanente has shown me that small steps can cover large distances. Of course we still have a long way to go. Any health-care reform for the future must have at its foundation access to affordable food that's good for the people who eat it, good for the people who grow it, and good for the planet.

---

Since starting the market in 2003, I've been offering a recipe of the week that started out going to an email distribution list to just my hospital. Before I knew it, 11,000 people had signed up to receive it each week. I felt that people might not be intimidated by my recipes on the theory that if I, an untrained chef, could do it, anyone could do it. We changed to a blog format in September 2007, and I receive many useful comments and suggestions from readers. Check it out at kp.org/farmersmarketrecipes.

Here are a few postings to take with you to your local farmers' market. Enjoy them in good health.

## EGGPLANT, TOMATO, AND ONION GRATIN

### SERVES 6

These vegetables are often found in baskets, side by side, at your local farmers' market in the late summer and early fall. Together with other fresh vegetables, they make a rich and piquant pasta sauce. I found a great recipe, and then modified it to make it just a little easier for basic cooks like me.

**Ingredients:**
1 pound Japanese or other elongated eggplants
1/2 onion, chopped

1 carrot, diced

1 stalk of celery, diced

Large pinch or two of crushed red chilies

4 cloves garlic, minced

1/3 bunch parsley, chopped

1 1/2 pounds plum tomatoes, cherry tomatoes, or heirlooms, diced

A dozen large basil leaves, coarsely sliced

1 tablespoon capers, rinsed and drained

8–12 kalamata olives, rinsed, drained, and coarsely chopped

4 tablespoons extra virgin olive oil

Salt and freshly ground pepper to taste

12 ounces whole wheat pasta in the shape of your choice

Preheat the oven to 425 degrees. Cut the eggplant into chunks, 1″ or less. Roast on a heavy baking sheet for about 45 minutes, or until browned and completely softened. Alternatively, you can roast them whole, having pricked them with a fork, cutting them in half when cool, and scraping out the flesh. The first method is easier, and I kind of like the skin, but then I even eat hairy kiwis whole.

Meanwhile, heat the olive oil and sauté the onion, carrot, and celery until soft. Mix in the crushed chilies, garlic, and parsley. Sauté until fragrant, then add the tomatoes. Cook them until you have a chunky sauce, seasoning to taste with salt and pepper. Add a little water to keep it saucy. Stir in the chunks of eggplant and basil. Simmer for a few minutes, and then stir in the capers and chopped olives.

Cook the pasta. Toss it with the sauce. Garnish with a little grated cheese or more parsley. This is a winner.

| Nutrition information per serving: | Fat: 11 gm |
|---|---|
| | Saturated fat: 1 gm |
| | Trans fat: 0 gm |
| Calories: 340 | Cholesterol: 0 mg |
| | Carbohydrate: 56 gm |
| | Fiber: 9 gm |
| | Sodium: 228 mg |
| | Protein: 10 gm |

## CAULIFLOWER WITH RED ONION, CHILIES, AND GARLIC

### SERVES 6

I have a somewhat checkered past when it comes to cauliflower. In my childhood, frozen bricks of the vegetable were boiled for hours and then topped with Velveeta cheese sauce—with pimentos, if we had company. Things have looked up since. I have learned to mostly like it steamed, in a stir fry, in gratins (bread crumbs help), and even in raw chunks on those ubiquitous vegetable platters served with an unknown white creamy dip at offsite meetings.

This cauliflower recipe is my favorite of all. It also works with broccoli, or a mix of cauliflower and broccoli.

**Ingredients:**

3 small heads of cauliflower or 1 large head, cut into ½" florets

⅓ large red onion, thinly sliced

2 large pinches crushed red chili flakes

4 cloves garlic, minced

Juice of ½ small lemon

2 splashes of red wine vinegar

Salt and freshly ground pepper to taste

3 tablespoons olive oil

Heat the olive oil in a skillet. Sauté the cauliflower until it starts to soften. Add the red onion and chilies. Continue to cook until the cauliflower starts to brown a little. Remove from the heat. Toss with the garlic. Splash it with the lemon juice and red wine vinegar. Season with salt and pepper and serve. This is a wonderfully simple and vibrant dish. Try it.

| Nutrition information per serving:<br><br>Calories: 100 | Fat: 7 gm |
|---|---|
| | Saturated fat: 1 gm |
| | Trans fat: 0 gm |
| | Cholesterol: 0 mg |
| | Carbohydrate: 9 gm |
| | Fiber: 4 gm |
| | Sodium: 137 mg |
| | Protein: 3 gm |

## GREEN BEANS AND TOMATOES

### SERVES 4

Many people talk about the health-related and obesity problems based on our diets and are particularly concerned about today's children. There's hope. A patient of mine has children ages twenty, thirteen, and nine. It's the nine-year-old girl that motivates the rest of the family. She bargains with her older brother to trade her meat to get his vegetables even though her Mom makes enough veggies for everyone.

Here's a recipe I think would be a good one for the girl to make for the whole family based on the produce I found at my farmers' market one Friday.

**Ingredients:**
> Three big handfuls green beans, stem end broken off, and cut in half
> 1 pint cherry tomatoes, halved
> 2 tablespoons olive oil
> 4–5 cloves garlic, coarsely minced
> 1 teaspoon dried oregano
> 2 ounces parmesan cheese, grated
> Salt and freshly ground pepper to taste

Preheat oven to 325 degrees. Steam the green beans for about 3 minutes in a saucepan on the stove. Heat the olive oil in a medium skillet and sauté the garlic for a minute. Add the tomatoes, green beans, and oregano. Cook for another couple of minutes. Season with salt and pepper. Transfer the veggies to a small baking dish. Sprinkle with the grated parmesan cheese. Bake about 15 minutes. Serve hot.

| Nutrition information per serving: | Fat: 11 gm |
|---|---|
| | Saturated fat: 3 gm |
| | Trans fat: 0 gm |
| Calories: 172 | Cholesterol: 12 mg |
| | Carbohydrate: 12 gm |
| | Fiber: 5 gm |
| | Sodium: 225 mg |
| | Protein: 8 gm |

# TO LEARN MORE—BOOKS, WEBSITES, AND ORGANIZATIONS OFFERING FURTHER INSIGHT INTO AMERICA'S FOOD SYSTEM AND ITS FUTURE

## BOOKS

Baur, Gene. *Farm Sanctuary: Changing Hearts and Minds About Animals and Food.* New York: Touchstone, 2008.

President and cofounder of Farm Sanctuary, the nation's leading farm animal protection organization, Baur eloquently advocates the fair treatment of farm animals, a return to the roots of agriculture, and the end of animals' suffering for our consumption.

Berry, Wendell. *What Are People For?* New York: North Point Press, 1990.

A collection of brilliant and thought-provoking essays by the novelist, poet, cultural critic, and farmer whom many consider the greatest living advocate for traditional, sustainable rural lifestyles.

———. *The Unsettling of America: Culture and Agriculture.* San Francisco: Sierra Club Books, 2004.

Originally published in 1977, this is Berry's best-known and most influential book, described by *Publishers Weekly* as "a cool, reasoned, lucid and at times poetic explanation of what agribusiness and the mechanization of farming are doing to the American fabric."

———. *The Way of Ignorance.* Berkeley: Counterpoint, 2005.

Berry's most recent collection of essays, *The Way of Ignorance* recommends the value of recognizing and accepting the limits of human knowledge and power, rather than striving, as we have, to remake society and even nature itself to conform with our fantasies of control.

Bryce, Robert. *Gusher of Lies: The Dangerous Delusions of Energy Independence.* New York: PublicAffairs, 2008.

Bryce explains why the idea of "energy independence" appeals to voters while also showing that renewable sources such as wind and solar are unlikely to meet America's growing energy demands. Along the way, Bryce eviscerates the ethanol "scam," which he calls one of the longest-running robberies ever perpetrated on American taxpayers.

Goodall, Jane. *Harvest for Hope: A Guide to Mindful Eating.* New York: Wellness Central/Hachette, 2006.

The renowned primatologist offers her personal appeal on behalf of more sustainable habits of eating and otherwise consuming the world's limited resources. Goodall explains the important benefits of local eating, vegetarianism, and even such simple conservation steps as turning off the water while you brush your teeth.

Heron, Katrina, ed. *Slow Food Nation's Come to the Table: The Slow Food Way of Living.* San Francisco: Modern Times, 2008.

Alice Waters, the celebrated chef and food activist, introduces a remarkable group of resilient fresh-food artisans who are committed to keeping our food supply delicious, diverse, and safe—for humans and the planet. Explore local flavors, wit, and wisdom along with the universal values of a food system that is "good, clean, and fair."

Hirshberg, Gary. *Stirring It Up: How to Make Money and Save the World.* New York: Hyperion, 2008.

Founder and CEO of Stonyfield, Hirshberg calls on individuals to realize their power to effect change in the marketplace—"the power of one"—while showing that environmental commitment makes for a healthier planet and a healthier bottom line, drawing from his twenty-five years of business experience as well as the examples of like-minded companies, such as Newman's Own, Patagonia, and Timberland.

Kimbrell, Andrew, ed. *Fatal Harvest: The Tragedy of Industrial Agriculture.* Sausalito, California: Foundation for Deep Ecology, 2002.

An unprecedented look at our current ecologically destructive agricultural system and a compelling vision for an organic and environmentally safer way of producing the food we eat, including more than 250 photographs and more than forty essays by leading ecological thinkers, including Wendell Berry, Wes Jackson, David Ehrenfeld, Helena Norberg-Hodge, Vandana Shiva, and Gary Nabhan.

———. *The Fatal Harvest Reader.* Washington, D.C.: Island Press, 2002.

Brings together in an affordable paperback edition the essays included in *Fatal Harvest*, offering a concise overview of the failings of industrial agriculture and approaches to creating a more healthful and sustainable food system.

Kingsolver, Barbara, with Steven L. Hopp and Camille Kingsolver. *Animal, Vegetable, Miracle: A Year of Food Life.* New York: Harper Collins, 2007.

Part memoir, part journalistic investigation, this book tells the story of how one family was changed by one year of deliberately eating food produced in the place where they live. Includes nutritional information, meal plans, and recipes.

Lappé, Anna, and Bryant Terry. *Grub: Ideas for an Urban Organic Kitchen.* New York: Jeremy P. Tarcher, 2006.

Combining a straight-to-the-point exposé about organic foods and the how-to's of creating an affordable, easy-to use organic kitchen, *Grub* brings organics home to urban-dwellers, giving the reader compelling arguments for buying organic food, revealing the pesticide industry's influence on government regulation, and the extent of its pollution in our waterways and bodies. Includes recipes, resource lists, tip sheets, charts, and checklists.

Lappé, Frances Moore, and Anna Lappé. *Hope's Edge: The Next Diet for a Small Planet.* New York: Jeremy P. Tarcher, 2003.

Describing their journeys through Brazil, Pakistan, Holland and the United States, the Lappés (mother and daughter) question the economic status quo as well as discuss the way different countries handle food production in times of scarcity and plenty. What we eat directly, they argue, connects us to the Earth and people around the globe.

McKibben, Bill. *Deep Economy: The Wealth of Communities and the Durable Future.* New York: Henry Holt, 2007.

Makes the case for moving beyond "growth" as the paramount economic ideal and pursuing prosperity locally, with regions producing more of their own food, generating more of their own energy, and creating more of their own culture and entertainment. Our purchases need not be at odds with the things we truly value, McKibben argues, and the more we nurture the essential humanity of our economy, the more we will recapture our own.

Nestle, Marion. *Food Politics: How the Food Industry Influences Nutrition.* Berkeley: University of California Press, 2003.

A brilliant exposé and analysis of how agribusiness lobbyists, corporate political donors, and ham-fisted regulators have turned government farm subsidies and nutrition policies into a massive welfare program for some of America's richest companies.

———. *Safe Food: Biology, Biotechnology, and Bioterrorism.* Berkeley: University of California Press, 2004.

Nestle argues that ensuring safe food involves not just washing hands or cooking food to higher temperatures but also politics. When it comes to food safety, industry, government, and consumers collide over issues of values, economics, and political power—and not always in the public interest. She demonstrates how powerful food industries oppose safety regulations, deny accountability, and blame consumers when something goes wrong, and how century-old laws for ensuring food safety no longer protect our food supply.

————. *What to Eat.* New York: North Point Press, 2007.

How we choose which foods to eat is growing more complicated by the day, and *What to Eat* offers a straightforward, practical approach to this dilemma. As Nestle takes us through each supermarket section—produce, dairy, meat, fish—she explains the issues, cutting through jargon and complicated nutrition labels and debunking the misleading health claims made by big food companies to help readers make wise food choices and eat sensibly and nutritiously.

Patel, Raj. *Stuffed and Starved: The Hidden Battle for the World Food System.* Hoboken, New Jersey: Melville House, 2008.

Half the world is malnourished, the other half obese—both symptoms of the corporate food monopoly. Raj Patel conducts a global investigation, traveling from the "green deserts" of Brazil and protester-packed streets of South Korea to bankrupt Ugandan coffee farms and barren fields in India, to explain the steps to regain control of the global food economy, stop the exploitation of farmers and consumers, and rebalance global sustenance.

Petrini, Carlo. *Slow Food: The Case for Taste.* New York: Columbia University Press, 2003.

The founder of the Slow Food movement, which seeks to revolutionize the way Americans shop for groceries, prepare and consume their meals, and think about food, recalls the movement's origins, first steps, and international expansion. *Slow Food* is also a powerful expression of the organization's goal of engendering social reform through the transformation of our attitudes about food and eating.

Pollan, Michael. *The Omnivore's Dilemma: A Natural History of Four Meals.* New York: Penguin, 2006.

Pollan uses the stories of four meals, their contents raised and prepared in three ways (industrially, pastorally, and personally—by Pollan himself), to illustrate the web of economic, social, political, and cultural relationships that determine what and how we eat.

————. *In Defense of Food: An Eater's Manifesto.* New York: Penguin, 2008.

Pollan assails "nutritionism," the pseudoscientific approach to eating that attempts to break down foods into their chemical components rather than viewing them as integral parts of a chain of natural processes. In its place, Pollan proposes a simple but liberating rule for life: "Eat food. Not too much. Mostly plants."

Pringle, Peter. *Food, Inc.: Mendel to Monsanto—The Promises and Perils of the Biotech Harvest.* New York: Simon & Schuster, 2003.

In the war over genetically modified foods, a handful of corporate giants, such as Monsanto, are pitted against a worldwide network of anticorporate ecowarriors like Greenpeace. Pringle suggests that a partnership among consumers, corporations, scientists, and farmers could still allow the biotech harvest to reach its full potential in helping to overcome the problem of world hunger, providing nutritious food, and keeping the environment healthy.

Roberts, Paul. *The End of Food*. Boston: Houghton Mifflin, 2008.

Journalist Roberts analyzes the shortcomings of the worldwide industrial food system and shows how our current methods of making, marketing, and moving what we eat are growing less and less compatible with the needs of the billions of consumers the system was supposedly built to serve.

Salatin, Joel. *Holy Cows and Hog Heaven: The Food Buyer's Guide to Farm-Fresh Food*. Swoope, Virginia: Polyface, Inc., 2006.

Ecological farmer Salatin offers advice and information designed to empower food buyers to pursue positive alternatives to the industrialized food system, learn about sustainable food production methods, and create a food system that enhances nature's ecology for future generations.

_____. *Everything I Want to Do Is Illegal: War Stories from the Local Food Front*. Swoope, Virginia: Polyface, Inc., 2007.

With humor and a sharp satiric touch, Salatin recounts his adventures in coping with a labyrinth of government regulations and misguided social perceptions that limit his ability to grow and sell foods raised the old-fashioned way, using natural, sustainable, traditional artisanal techniques.

Schlosser, Eric. *Fast Food Nation: The Dark Side of the All-American Meal*. New York: Houghton Mifflin, 2001.

Schlosser steps behind the counter of McDonald's, KFC, and Burger King to analyze how the rise and dominance of fast food has transformed the American food industry, with dire consequences for the health of consumers, the rights of workers, and the safety of our food supply.

Shiva, Vandana. *Stolen Harvest: The Hijacking of the Global Food Supply*. Cambridge, Mass.: South End Press, 2000.

Indian scientist and activist Shiva charts the impacts of globalized, corporate agriculture on small farmers, the environment, and the quality of the food we eat. She shows that, when the food system is industrialized, millions of peasants are forced off the land, and a system of agriculture that was once ecologically friendly and diverse is replaced by monoculture cultivation that can only be supported by toxic chemicals.

_____, ed. *Manifestos on the Future of Food and Seed*. Cambridge, Mass.: South End Press, 2007.

A collection of essays by writers including Carlo Petrini, Michael Pollan, and Prince Charles. Dealing with such questions as: How are seeds cultivated and saved? How far must food travel before reaching our plate? Who gets paid for the food we eat? And, why does our food taste like this?, *Manifestos* offers ideas for building a more sustainable system for producing and distributing the world's food supply.

————. *Soil Not Oil: Environmental Justice in an Age of Climate Crisis.* Cambridge, Mass.: South End Press, 2008.

Connecting the dots between industrial agriculture and climate change, Shiva shows that a world beyond dependence on fossil fuels and globalization is both possible and necessary. Her solution: an agriculture based on self-organization, sustainability, and community rather than corporate power and profits.

Singer, Peter. *Animal Liberation.* New York: Harper Perennial, 2001.

Originally published in 1975, this landmark book is widely credited with launching the modern animal rights movement. A professor of bioethics at Princeton University and one of today's leading philosophers, Singer explores the moral bankruptcy not just of factory farming but also of animal experimentation, product testing, and other widespread practices that abuse and torture living things for short-term economic benefit.

Taubes, Gary. *Good Calories, Bad Calories.* New York: Knopf, 2007.

For decades we have been taught that fat is bad for us, carbohydrates better, and that the key to a healthy weight is eating less and exercising more. Yet despite this advice, we have seen unprecedented epidemics of obesity and diabetes. Award-winning science writer Taubes argues that the problem lies in refined carbohydrates, like white flour, easily digested starches, and sugars, and that the key to good health is the kind of calories we take in, not the number.

Winne, Mark. *Closing the Food Gap: Resetting the Table in the Land of Plenty.* Boston: Beacon Press, 2008.

Winne tells the story of how we get our food, from poor people at food pantries or bodegas and convenience stores to the more comfortable classes, who increasingly seek out organic and local products. Winne shows how communities since the 1960s have responded to malnutrition with a slew of strategies and methods, all against a backdrop of ever-growing American food affluence and gastronomical expectations.

Yunus, Muhammad, with Karl Weber. *Creating a World Without Poverty: Social Business and the Future of Capitalism.* New York: PublicAffairs, 2007.

The pioneer of microcredit and cowinner of the 2006 Nobel Peace Prize describes how the current economic system creates poverty and explains how "social business" can create just, sustainable economic growth for the world's billions of poor people.

## WEBSITES AND ORGANIZATIONS

### Alliance for a Healthier Generation

http://www.healthiergeneration.org

A partnership between the American Heart Association and the William J. Clinton Foundation that seeks to address one of the nation's leading public health threats—childhood obesity. The goal of the alliance is to reduce the nationwide prevalence of childhood obesity by 2015 and to empower kids nationwide to make healthy lifestyle choices.

### American Community Gardening Association

http://www.communitygarden.org

A binational (U.S. and Canadian) nonprofit membership organization of professionals, volunteers, and supporters of community greening in urban and rural communities.

### American Corn Growers Association (ACGA)

http://www.acga.org

Founded in 1987, the American Corn Growers Association is America's leading progressive commodity association, representing the interests of corn producers in thirty-five states. ACGA has become a key player in the development of farm and trade policy along with production issues such as seed patent law, GMO policies, and the pursuit of renewable energy sources such as wind power.

### California Center for Public Health Advocacy

http://www.publichealthadvocacy.org

Raises awareness about critical public health issues and mobilizes communities to promote the establishment of effective state and local health policies.

### The Center for Ecoliteracy

http://www.ecoliteracy.org

Provides tools, ideas, and support for combining hands-on experience in the natural world with curricular innovation in K–12 education.

### Center for Foodborne Illness Research & Prevention

http://foodborneillness.org

Founded in 2006 to help America create innovative, science-based solutions for the food challenges of the twenty-first century, CFI works with federal, state, and local government, as well as farmers; food processors/distributors/retailers; medical providers; educators; policy-makers; and consumers to develop better food protections and ultimately improve public health through research, education, and advocacy.

### The Center for Food Safety
http://www.centerforfoodsafety.org

A nonprofit public interest and environmental advocacy membership organization established in 1997 by its sister organization, International Center for Technology Assessment, for the purpose of challenging harmful food production technologies and promoting sustainable alternatives. CFS combines multiple tools and strategies in pursuing its goals, including litigation and legal petitions for rulemaking; legal support for various sustainable agriculture and food safety constituencies; and public education, grassroots organizing, and media outreach.

### The Center for Science in the Public Interest
http://www.cspinet.org/about/index.html

Founded in 1971, CSPI is an advocate for nutrition and health, food safety, alcohol policy, and sound science. Its award-winning newsletter, *Nutrition Action Healthletter*, with some 900,000 subscribers in the United States and Canada, is the largest-circulation health newsletter in North America.

### Coalition of Immokalee Workers
http://www.ciw-online.org

The CIW is a community-based worker organization whose members are largely Latino, Haitian, and Mayan Indian immigrants working in low-wage jobs throughout the state of Florida, mostly employed by large agricultural corporations in the tomato and citrus harvests. CIW fights for, among other things: a fair wage, more respect from bosses and employers, better and cheaper housing, stronger laws and stronger enforcement against those who would violate workers' rights, the right to organize without fear of retaliation, and an end to indentured servitude in the fields.

### The Community Food Security Coalition
http://www.foodsecurity.org

A North American coalition of diverse people and organizations working from the local to international levels to build community food security. The coalition includes almost 300 organizations from social and economic justice, antihunger, environmental, community development, sustainable agriculture, community gardening, and other fields, and is dedicated to building strong, sustainable, local and regional food systems that ensure access to affordable, nutritious, and culturally appropriate food to all people at all times.

### Consumer Federation of America
http://www.consumerfed.org

A consumer advocacy lobbying organization whose members include some 300 nonprofit organizations from throughout the nation, with a combined membership exceeding fifty million people.

## The Cool Foods Campaign

http://coolfoodscampaign.org

A project of the Center for Food Safety and the CornerStone Campaign, the Cool Foods Campaign makes the connections between the foods we eat and their contribution to global warming, aiming to educate the public about the impact of their food choices across the entire food system and to empower them with the resources to reduce this impact.

## Council for Responsible Genetics

http://www.gene-watch.org

Founded in 1983, the Council for Responsible Genetics fosters public debate about the social, ethical, and environmental implications of genetic technologies.

CRG also publishes a bimonthly magazine, *GeneWatch*, the only publication of its kind in the nation.

## Ecological Farming Association

http://www.eco-farm.org

Dedicated to the development of ecologically based food systems, both domestically and throughout the world, by educating farmers, the agriculture industry, and other stewards of the land about practical and economically viable techniques of ecological agriculture, informing consumers and policy-makers about ecological food production and its connection to the health of people and communities. EFA promotes alliances between individuals and organizations who share its vision of a transformed global food system.

## FactoryFarm.org

http://www.factoryfarm.org

Believing that factory farms destroy the environment, threaten human health, devastate local communities, and compromise animal welfare, Factory-Farm.org works to educate the public about these problems in order to help promote healthy, sustainable alternatives to industrial agriculture.

## Farm to School

http://www.farmtoschool.org

Farm to School brings healthy food from local farms to schoolchildren nationwide. Its program teaches students about the path from farm to fork and instills healthy eating habits that can last a lifetime. At the same time, use of local produce in school meals and educational activities provides a new direct market for farmers in the area and mitigates the environmental impacts of transporting food long distances.

**Feeding America**

http://feedingamerica.org

Formerly known as Second Harvest, Feeding America is a network of individuals, local food banks, national offices, and corporate and government partners working to secure food and grocery products on a national level to distribute to local food banks. Feeding America also helps set standards for food safety, financial systems, and record-keeping among local food banks as well as transportation and donor relations.

**FoodFirst Information and Action Network (FIAN)**

http://www.fian.org

An international human rights organization focusing on the global food crisis.

**Food & Water Watch**

http://www.foodandwaterwatch.org

A nonprofit consumer organization that works to ensure clean water and safe food in the United States and around the world, challenging the corporate control and abuse of food and water resources by empowering people to take action and by transforming the public consciousness about what we eat and drink.

**The Food Trust**

http://www.thefoodtrust.org

Founded in 1992, the Food Trust works to improve the health of children and adults, promote good nutrition, increase access to nutritious foods, and advocate for better public policy. The Food Trust runs a farmers' market program; school food programs serving more than one hundred public schools in southeast Pennsylvania; and the Kindergarten Initiative, a multifaceted nutrition program that connects children and parents to locally grown food, and serves as a partner in one of the fifteen most innovative government programs in the nation, The Fresh Food Financing Initiative.

**Heifer International**

http://www.heifer.org

Envisioning a world of communities living together in peace and equitably sharing the resources of a healthy planet, Heifer International works to end hunger and poverty and to care for the Earth by giving families a source of food rather than short-term relief. Over sixty years, millions of families in 128 countries have been given the gifts of self-reliance and hope through Heifer programs.

### The Humane Society of the United States
http://www.hsus.org

Established in 1954, the HSUS is the nation's largest and most effective animal protection organization, seeking a humane and sustainable world for all animals—a world that will also benefit people. In particular, the HSUS takes a leadership role on farm animal advocacy issues, working to end intensive confinement, mutilations without painkillers, and inhumane slaughter practices.

### The Institute for Food and Development Policy/Food First
http://www.foodfirst.org

The Institute for Food and Development Policy/Food First works to analyze the root causes of global hunger, poverty, and ecological degradation and develop solutions in partnership with movements working for social change.

### The Land Institute
http://www.landinstitute.org

The Land Institute has worked for more than twenty years on the problem of agriculture, seeking to develop an agricultural system with the ecological stability of the prairie and a grain yield comparable to that from annual crops—an approach it calls Natural Systems Agriculture. Institute experts have researched, published in refereed scientific journals, and given hundreds of public presentations here and abroad.

### Local Harvest
http://www.localharvest.org

America's leading organic and local food website, Local Harvest maintains a definitive and reliable public nationwide directory of small farms, farmers' markets, and other local food sources. The site's search engine helps people find products from family farms and local sources of sustainably grown food and encourages them to establish direct contact with small farms in their local area. Its online store helps small farms develop markets for some of their products beyond their local area.

### National Campaign for Sustainable Agriculture
http://www.sustainableagriculture.net

A nationwide partnership of diverse individuals and organizations cultivating grassroots efforts to engage in policy development processes that result in food and agricultural systems and rural communities that are healthy, environmentally sound, profitable, humane, and just.

## Organic Consumers Association
http://www.organicconsumers.org

OCA is an online and grassroots organization campaigning for health, justice, and sustainability, and focusing on the crucial issues of food safety, industrial agriculture, genetic engineering, children's health, corporate accountability, fair-trade, environmental sustainability, and other key topics. The only organization in the United States focused exclusively on promoting the views and interests of the nation's estimated fifty million organic and socially responsible consumers, the OCA also represents more than 850,000 members, subscribers, and volunteers, including several thousand businesses in the natural foods and organic marketplace.

## Oxfam International
http://www.oxfam.org

In partnership with more than 3,000 local organizations, Oxfam works with people living in poverty striving to exercise their human rights, assert their dignity as full citizens, and take control of their lives.

## Participant Media, Inc.
http://www.participantmedia.com

Participant Media is the leading provider of entertainment that inspires and compels social change. Whether it is a feature film, documentary, or other form of media, Participant exists to tell compelling, entertaining stories that also create awareness of the real issues that shape our lives. You can learn more about the movie *Food, Inc.* by visiting Participant's website.

## Pesticide Action Network North America (PANNA)
http://www.panna.org

PANNA works to replace pesticide use with ecologically sound and socially just alternatives. As one of five autonomous PAN Regional Centers worldwide, PANNA links local and international consumer, labor, health, environment, and agriculture groups into an international citizens' action network that challenges the global proliferation of pesticides, defends basic rights to health and environmental quality, and works to ensure the transition to a just and viable society.

## Pew Commission on Industrial Farm Animal Production (PCIFAP)
http://www.ncifap.org/index.html

Funded by the Pew Charitable Trusts and the Johns Hopkins Bloomberg School of Public Health, PCIFAP was formed to conduct a comprehensive, fact-based, and balanced examination of key aspects of the farm animal industry. Commissioners represent diverse backgrounds and perspectives and come from the fields of veterinary medicine, agriculture, public health, business, government, rural advocacy, and animal welfare. The PCIFAP website is a useful source of scientifically based information on how the farm production industry influences public health, farm communities, and the welfare of animals.

**Polyface Farms**
http://www.polyfacefarms.com
Polyface, Inc., is a family-owned, multigenerational, pasture-based, beyond organic, local-market farm and informational outreach in Virginia's Shenandoah Valley. In addition to producing foods such as Salad Bar Beef, Pigaerator Pork, and Pastured Poultry, Polyface is in the redemption business: healing the land, healing the food, healing the economy, and healing the culture.

**The Robert Wood Johnson Foundation**
http://www.rwjf.org
The Robert Wood Johnson Foundation focuses on improving both the *health* of everyone in America and their health *care*—how it's delivered, how it's paid for, and how well it serves patients and their families.

**Seed Savers Exchange**
http://www.seedsavers.org
Founded in 1975, Seed Savers Exchange is a nonprofit organization that saves and shares heirloom seeds of fruits, vegetables, flowers, and herbs. Members of the organization have exchanged about one million samples of rare garden seeds, forming a living legacy that helps to nurture our diverse, fragile genetic and cultural heritage.

**Slow Food USA**
http://www.slowfoodusa.org
Slow Food is an idea, a way of living, and a way of eating. It is also a global, grassroots movement with thousands of members around the world that links the pleasure of food with a commitment to community and the environment. Slow Food USA works to inspire a transformation in food policy, production practices, and market forces so that they ensure equity, sustainability, and pleasure in the food Americans eat.

**The Small Planet Institute**
http://www.smallplanet.org
Frances Moore Lappé and Anna Lappé founded the Small Planet Institute in 2001 to help define, articulate, and further the worldwide shift from the dominant, failing notion of democracy as a set of fixed institutions toward democracy understood as a way of life, a culture in which the values of inclusion,

fairness, and mutual accountability infuse all dimensions of public life. The Institute seeks to further this historic transition through collaborative public education efforts with colleagues worldwide and through books, articles, speeches, and other media.

### Sustainable Agriculture Research and Education
http://www.sare.org

Since 1988, the Sustainable Agriculture Research and Education (SARE) program has helped advance farming systems that are profitable, environmentally sound, and good for communities through a nationwide research and education grants program.

### Sustainable Table
http://www.sustainabletable.org

Created in 2003 by the nonprofit organization GRACE, Sustainable Table celebrates local sustainable food, educates consumers on food-related issues, and works to build community through food. Its website is home to the Eat Well Guide, an online directory of sustainable products in the United States and Canada, and the critically acclaimed, award-winning Meatrix movies, *The Meatrix*, *The Meatrix II: Revolting*, and *The Meatrix II½*.

### TransFair USA and Fair Trade Certified
http://www.transfairusa.org

TransFair USA and Fair Trade Certification empower farmers and farmworkers to lift themselves out of poverty by investing in their farms and communities, protecting the environment, and developing the business skills necessary to compete in the global marketplace. TransFair USA is the only independent, third-party certifier of fair-trade products in the United States and one of twenty members of Fairtrade Labelling Organizations International (FLO). Fair Trade Certification is currently available in the United States for coffee, tea and herbs, cocoa and chocolate, fresh fruit, sugar, rice, and vanilla.

### United Farm Workers
http://www.ufw.org

Founded in 1962 by Cesar Chavez, the United Farm Workers of America is the nation's first successful and largest farmworkers' union, currently active in ten states. Recent years have witnessed dozens of key UFW union contract victories, among them the largest strawberry, rose, winery, and mushroom firms in California and the nation. Many recent UFW-sponsored laws and regulations aid farmworkers; in California, the first state regulation in the United States prevents further heat deaths of farmworkers. The UFW is also pushing its historic bipartisan and broadly backed AgJobs immigration reform bill.

**United Food and Commercial Workers International Union**
http://www.ufcw.org
UFCW's 1.3 million members work in a range of industries, with the majority working in retail food, meatpacking and poultry, food processing and manufacturing, and retail stores. The UFCW is also the largest union of young workers, with forty percent of UFCW members under the age of thirty.

**World Hunger Year**
http://www.worldhungeryear.org
Founded in 1975, WHY is a leader in the fight against hunger and poverty in the United States and around the world. WHY advances long-term solutions to hunger and poverty by supporting community-based organizations that empower individuals and build self-reliance by offering job training, education, and after-school programs; increasing access to housing and health care; providing microcredit and entrepreneurial opportunities; teaching people to grow their own food; and assisting small farmers.

# NOTES

## ANOTHER TAKE: FOOD SAFETY
## CONSEQUENCES OF FACTORY FARMS

1. Union of Concerned Scientists, "Hogging It!: Estimates of Antibiotic Abuse in Livestock," *UCS*, 2001.

2. Union of Concerned Scientists, "Food and Environment: Antibiotic Resistance," *UCS*, October 2003.

3. Keep Antibiotics Working, "The Health Threat," www.keepantibioticsworking.com.

4. American Medical Association, "Antibiotics and Antimicrobials," http://www.ama-assn.org/ama/pub/category/1863.html; National Institute of Allergy and Infectious Disease, "The Problem of Antimicrobial Resistance," http://www.niaid.nih.gov/factsheets/antimicro.htm, April 2006; American Public Health Association, "Antibiotic Resistance Fact Sheet," http://www.apha.org/advocacy/reports/facts/advocacyfactantibiotic.htm.

5. Scott A. McEwen and Paula J. Fedorka-Cray, "Antimicrobial Use and Resistance in Animals," *Clinical Infectious Diseases* 34, suppl. 3 (2002): S93–106.

6. Francisco Diez-Gonzalez, Todd R. Callaway, Menas G. Kizoulis, and James B. Russell, "Grain Feeding and the Dissemination of Acid-Resistant Escherichia coli from Cattle," *Science* 281, no. 5383 (September 11, 1998): 1666–1668.

7. Union of Concerned Scientists, "Greener Pastures: How Grass-Fed Beef and Milk Contribute to Healthy Eating," 2006.

8. Janet Raloff, "Hormones: Here's the Beef: Environmental Concerns Reemerge over Steroids Given to Livestock," *Science News* 161, no. 1 (January 5, 2002): 10.

9. The Scientific Committee on Veterinary Measures Relating to Public Health, "Assessment of Potential Risks to Human Health from Hormone Residues in Bovine Meat and Meat Products," European Commission, April 30, 1999.

10. Ibid.

11. "Bovine Somatotropin (bST)," Biotechnology Information Series (Bio-3), North Central Regional Extension Publication, Iowa State University—University Extension, December 1993.

12. Susan M. Cruzan, FDA press release on rBST approval. Food and Drug Administration. November 5, 1993.

13. APHIS, "Bovine Somatotropin: Info Sheet," USDA, May 2003.

14. I. Doohoo et al, "Report of the Canadian Veterinary Medical Association Expert Panel on rBST" (Executive Summary), *Health Canada*, November 1998.

15. S. S. Epstein, "Unlabeled Milk from Cows Treated with Biosynthetic Growth Hormones: A Case of Regulatory Abdication," *International Journal of Health Services* 26, no. 1 (1996): 173–185.

16. G. Steinman, "Can the Chance of Having Twins Be Modified by Diet?" *Lancet*, 367, no. 9521 (May 6, 2006): 1461–1462.

17. "Fowl Play: The Poultry Industry's Central Role in the Bird Flu Crisis," *GRAIN*, February 2006, 2.

18. Jo Revill, "Turkey Carcasses from Hungary Linked to UK Bird Flu Outbreak," *The Observer*, February 8, 2007, www.observer.co.uk.

19. "Risk Management Evaluation for Concentrated Animal Feeding Operations," US Environmental Protection Agency, National Risk Management Laboratory, May 2004, 7.

## CHAPTER FIVE: THE ETHANOL SCAM

1. Lester R. Brown, "Starving the People to Feed the Cars," *Washington Post*, September 10, 2006, B3. Available online at http://www.washingtonpost.com/wp-dyn/content/article/2006/09/08/AR2006090801596_pf.html. For more on Brown's group, see http://www.earth-policy.org.

2. The U.S. Department of Agriculture estimates that distillers can get 2.7 gallons of ethanol per bushel of corn. In 2006, U.S. farmers produced about 10.5 billion bushels of corn. Converting all of that corn into ethanol would produce about 28.3 billion gallons of ethanol. However, ethanol's lower heat content means that the actual output would be equivalent to 18.7 billion gallons of gasoline, or about 1.2 million barrels per day. The United States now consumes about 21 million barrels of oil per day. Available online at http://www.eia.doe.gov/neic/quickfacts/quickoil.html.

3. Farmers can produce about 40 bushels of soybeans per acre, enough to make about 60 gallons of biodiesel. In 2006, U.S. farmers produced 3.188 billion bushels of soybeans. That quantity would yield about 4.8 billion gallons of diesel fuel per year or about 313,000 barrels of oil per day. For soybean production figures, see USDA data, available online at http://www.ers.usda.gov/News/soybeancoverage.htm.

4. If you assume that each of those 200-plus plants cost $75 million to construct (a conservative estimate), the total cost of those distilleries is about $15 billion. For reference on plant costs, VeraSun, a major ethanol producer, says in a recent annual report that one of its newest ethanol plants, a 110-million-gallon-per-year facility in Hartley, Iowa, cost about $66 million to construct. See http://www.sec.gov/Archives/edgar/data/1343202/000119312508053294/d10k.htm.

Another plant, an 84-million-gallon plant in Cloverdale, Indiana, owned by AltraBiofuels, opened in summer 2008. It cost $170 million. See http://earth2tech

.com/2008/05/19/altrabiofuels-names-new-ceo-starts-second-plant-production/ http://www.npr.org/templates/story/story.php?storyId=92559699.

5. Renewable Fuels Association, "Ethanol Biorefinery Locations," undated. Accessed Sept. 8, 2008. Available online at http://www.ethanolrfa.org/industry/locations/.

6. In August 2008, the USDA estimated that American corn production for the year would total 12.3 billion bushels, of which 4.1 billion bushels would be used for ethanol production. See Scott Kilman, "Bumper Harvests Not Enough to Ease Food Costs," *Wall Street Journal*, August 13, 2008, A3. Available online at http://online.wsj .com/article/SB121854537937633263.html?mod=hps_us_whats_news. In 2000, corn ethanol producers used about 571 million bushels. For data, see the Earth Policy Institute, which reports that in 2000, U.S. corn ethanol consumed 16 million tons of corn. With 35.7 bushels per ton, that equals 571.2 million bushels. Earth Policy data available online at http://www.earth-policy.org/Updates/2007/Update63_data2.htm#fig5.

7. U.S. Grains Council data. Available at http://www.grains.org/page.ww?section =Barley%2C+Corn+%26+Sorghum&name=Corn.

8. Brown, B3.

9. Environmental Working Group data. Available online at http://farm.ewg.org/ farm/region.php?fips=00000#topprogs.

10. USDA, "Global Agricultural Supply and Demand: Factors Contributing to the Recent Increase in Food Commodity Prices," revised July 2008, 6, 14. Available online at http://www.ers.usda.gov/Publications/WRS0801/WRS0801.pdf.

11. USDA, *Amber Waves*, September 2007, 39. Available online at http://www.ers .usda.gov/AmberWaves/September07/PDF/AW_September07.pdf.

12. USDA, "USDA Long-Term Projections, February 2007," 31. Available online at http://www.ers.usda.gov/publications/oce071/oce20071c.pdf.

13. Paul C. Westcott, "U.S. Ethanol Expansion Driving Changes Throughout the Agricultural Sector," U.S. Department of Agriculture, *Amber Waves*, September 2007, 13. Available online at http://www.ers.usda.gov/AmberWaves/September07/ PDF/AW_September07.pdf.

14. Ibid.

15. World population data from U.S. Census Bureau. Available online at http://www.census.gov/main/www/popclock.html. 2030 estimate from World Population Clock. Available online at http://www.worldometers.info/population/.

16. USDA, "Global Agricultural Supply and Demand: Factors Contributing to the Recent Increase in Food Commodity Prices," revised July 2008, 5.

17. Thomas Elam, "Biofuel Support Policy Costs to the U.S. Economy," The Coalition for Balanced Food and Fuel Policy, March 24, 2008, 25. Available online at http://www.balancedfoodandfuel.org/ht/a/GetDocumentAction/i/10560.

18. Donald Mitchell, "A Note on Rising Food Prices," World Bank, April 8, 2008, 1. Available online at http://image.guardian.co.uk/sys-files/Environment/documents/ 2008/07/10/Biofuels.PDF.

19. National Public Radio, "World Bank Chief: Biofuels Boosting Food Prices," April 11, 2008. Available online at http://www.npr.org/templates/story/story/php ?storyId=89545855.

20. Randy Schnepf, "High Agricultural Commodity Prices: What Are the Issues?" Congressional Research Service, updated May 29, 2008, 20. Available online at http://assets.opencrs.com/rpts/RL34474_20080529.pdf.

21. For more, see http://www.ifpri.org/about/about_menu.asp.

22. Mark W. Rosegrant, "Biofuels and Grain Prices: Impacts and Policy Responses," International Food Policy Research Institute, May 7, 2008. Available online at http://www.ifpri.org/pubs/testimony/rosegrant20080507.asp. Rosegrant's full quote:

> The increased biofuel demand during the period, compared with previous historical rates of growth, is estimated to have accounted for 30 percent of the increase in weighted average grain prices. Unsurprisingly, the biggest impact was on maize prices, for which increased biofuel demand is estimated to account for 39 percent of the increase in real prices. Increased biofuel demand is estimated to account for 21 percent of the increase in rice prices and 22 percent of the rise in wheat prices.

23. Ibid.

24. Ibid.

25. Renewable Fuels Agency, "The Gallagher Review of the Indirect Effects of Biofuels Production," 9. Available: http://www.dft.gov.uk/rfa/_db/_documents/Report_of_the_Gallagher_review.pdf.

26. USDA, "USDA Officials Briefing with Reporters on the Case for Food and Fuel," May 19, 2008. Available at http://www.usda.gov/wps/portal/%21ut/p/_s.7_0_A/7_0_1OB?contentidonly=true&contentid=2008/05/0130.xml.

27. USDA, Economic Research Service, "Food Security Assessment, 2007," July 2008, 1. Available online at www.ers.usda.gov/Publications/GFA19/GFA19.pdf.

28. National Corn Growers Association, "Understanding the Impact of Higher Corn Prices on Consumer Food Prices," updated April 18, 2007, 13. Available online at http://www.ncga.com/ethanol/pdfs/2007/FoodCornPrices.pdf.

29. USDA, "Food Security Assessment, 2007," 13.

30. Renewable Fuels Association, "Ethanol Facts: Food vs. Fuel," accessed September 8, 2008. Available online at http://www.ethanolrfa.org/resource/veetc/.

31. Grassley Watch, "Challenge Grassley on Ethanol, if You Agree with Him," May 20, 2008. Available online at http://www.grassleywatch.com/?p=19.

32. P. J. Crutzen, A. R. Mosier, K. A. Smith, and W. Winiwarter, "$N_2O$ Release from Agro-Biofuel Production Negates Global Warming Reduction by Replacing Fossil Fuels," Atmospheric Chemistry and Physics Discussions, August 1, 2007, 11191.

33. Government Accountability Office, "Tax Policy: Effects of the Alcohol Fuels Tax Incentives," March 1997, GAO/GGD-97-41, 6, 17.

34. Alex Farrell and Michael O'Hare, Memo to the California Air Resources Board, January 12, 2008, 3. Available online at http://www.arb.ca.gov/fuels/lcfs/011608ucb_luc.pdf.

35. Timothy Searchinger et al., "Use of U.S. Croplands for Biofuels Increases Greenhouse Gases Through Emissions from Land Use Changes," Science, February 7, 2008, 1238.

36. Environment News Service, "Conservationists Seek Firm Limits on Gulf Dead Zone Pollution," July 30, 2008. Available online at http://www.ens-newswire.com/ens/jul2008/2008-07-30-095.asp.

37. Reuters, "Gulf of Mexico 'Dead Zone' to Hit Record Size: NOAA," July 15, 2008. Available online at http://www.reuters.com/articlePrint?articleId=USN1533337 620080715.

38. U.S. Department of Energy, "Energy Demands on Water Resources," December 2006, 9. Available online at http://www.sandia.gov/energy-water/docs/121-Rpt ToCongress-EWwEIAcomments-FINAL.pdf.

39. Ibid., 61–62.

40. The math is straightforward: 1,000,000 Btus / 80,000 Btus per gallon of ethanol = 12.5 gallons of ethanol per MMBtu.

41. Environmental Protection Agency. "Exercise II. The Superior Car Wash," undated. Available online at http://www.epa.gov/nps/nps_edu/stopx2.htm.

42. David Pimentel and Tad W. Patzek, "Ethanol Production Using Corn, Switchgrass, and Wood: Biodiesel Production Using Soybean and Sunflower," *Natural Resources Research*, March 2005, 66. Available online at http://petroleum.berkeley.edu/ papers/Biofuels/NRRethanol.2005.pdf.

43. Nebraska Corn Board, "USDA Corn Production Report," September 15, 2006. Available online at http://www.nebraskacorn.org/cornmerch/usdareports.htm.

44. Fifteen percent of 885 is 132.75.

45. USDOE, 57, 59. Note that the report estimates that oil extraction requires between five and thirteen gallons of water per barrel of crude or, at most, 0.3 gallons of water per gallon of crude. Page 59 puts the refining requirements at "about 1 to 2.5 gallons" of water for each gallon of product.

46. Jan F. Kreider and Peter S. Curtiss, "Comprehensive Evaluation of Impacts from Potential, Future Automotive Fuel Replacements," *Proceedings of Energy Sustainability* 2007, June 27–30, 2007. Available online http://www.fuelsandenergy .com/papers/ES2007-36234.pdf.

47. Renewable Fuels Association data. For states, see
   http://www.ethanolrfa.org/industry/statistics/.
For numbers and capacity:
   http://www.ethanolrfa.org/industry/locations/

48. Renewable Fuels Association data, "VEETC," accessed September 8, 2008. Available online at http://www.ethanolrfa.org/resource/veetc/.

49. USDA data. Available online at http://www.nass.usda.gov/QuickStats/index2.jsp.

50. Environmental Working Group data. Available online at http://farm.ewg.org/ farm/progdetail.php?fips=00000&progcode=corn.

51. USDA data. Available online at http://www.nass.usda.gov/QuickStats/index2.jsp.

52. Environmental Working Group data. Available online at http://farm.ewg.org/ farm/progdetail.php?fips=00000&progcode=corn.

## ANOTHER TAKE: EXPOSURE TO PESTICIDES

1. MAFF, "Orchards and Fruit Stores in Great Britain 1996," Pesticide Usage Survey Report 142, (London: MAFF Publications, 1998); U.S. EPA, Office of the Administrator, "Environmental Health Threats to Children," EPA 175-F-96-001, September 1996; National Research Council, National Academy of Sciences, "Pesticides in the

Diets of Infants and Children," (Washington, D.C.: National Academy Press, 1993), 184–185.

2. U.S. EPA, "Draft Final Guidelines for Carcinogen Risk Assessment," EPA/630/P-03/001A, 2003. Accessed July 9, 2004. Available online at www.epa.gov/ncea/raf/cancer2003.htm.

3. Kenneth A. Cook et al., "Pesticides in the U.S. Food Supply," February 1995, www.ewg.org/reports/fruit/Contents.html; "CFSAN FDA Office of Plant and Dairy Foods: FDA Pesticide Residue Monitoring Program 1994–2002," http://vm.cfsan.fda.gov/~dms/pesrpts.html.

4. C. L. Curl, R. A. Fenske, and K. Elgethun, "Organophosphorus Pesticide Exposure of Urban and Suburban Pre-School Children with Organic and Conventional Diets," *Environmental Health Perspectives* 111, no. 3 (2003): 377–382.

5. Department of Health and Human Services, Centers for Disease Control, "National Report on Human Exposure to Environmental Chemicals," March 2003, www.cdc.gov/exposurereport/2nd; Saulk Institute, "Loss of Neuropathy Target Esterase in Mice Linking Organophosphate Exposure to Hyperactivity," *Nature Genetics*, March 2003, 477; Environmental Protection Agency, "America's Children and the Environment," March 2003, www.epa.gov/envirohealth/children.

6. DHHS, Centers for Disease Control.

7. W. P. Porter et al., "Groundwater Pesticides: Interactive Effects of Low Concentrations of Carbamates Aldicarb and Methomyl and the Triazine Metribuzin on Thyroxine and Somatrophin Levels in White Rats," *Journal of Toxicology and Environmental Health* 40 (1993): 15–34; C. A. Boyd et al., "Behavioral and Neurochemical Changes Associated with Chronic Exposure to Low Level Concentrations of Pesticide Mixtures," *Journal of Toxicology and Environmental Health* 30 (1990): 209–221; W. P. Porter et al., "Endocrine Immune and Behavioral Effects of Aldicarb (Carbamate), Atrazine (Triazine) and Nitrate (Fertilizer) Mixtures at Groundwater Concentrations," *Toxicology and Industrial Health* 15 (1999): 133–150; M. Thiruchelvam et al., "The Nigrostriatal Dopaminergic System as a Preferential Target of Repeated Exposures to Combined Paraquat and Maneb: Implications for Parkinson's Disease," *Journal of Neuroscience* 20, no. 24 (2000): 2907–2914; B. P. Baker et al., "Pesticide Data Program (2000–2002)," *Food Additives* 19, no.5 (2000): 427–446.

8. MAFF.

## CHAPTER SIX: THE CLIMATE CRISIS AT THE END OF OUR FORK

1. Henning Steinfeld et al., *Livestock's Long Shadow: Environmental Issues and Options* (Rome: Food and Agriculture Organization of the United Nations, 2006). While livestock is responsible for eighteen percent of total emissions, transportation is responsible for a total of thirteen percent of the global warming effect.

2. Film stats from Box Office Mojo. Available online at http://www.boxofficemojo.com/movies/?page=main&id=inconvenienttruth.htm.

3. R. A. Neff, I. L. Chan, and K. A. Smith, "Yesterday's Dinner, Tomorrow's Weather, Today's News?: US Newspaper Coverage of Food System Contributions to Climate Change," *Public Health Nutrition* (2008).

4. Rajendra Pachauri, "Global Warning—The Impact of Meat Production and Consumption on Climate Change," paper presented at the Compassion in World Farming, London, September 8, 2008.

5. Ibid.

6. N. H. Stern, *The Economics of Climate Change: The Stern Review* (Cambridge: Cambridge University Press, 2007), 539.

7. Ibid.

8. Ingredients for Quaker Granola Bar available online: https://www.weg-mans.com/webapp/wcs/stores/servlet/ProductDisplay?langId=1&storeId=10052&productId=359351&catalogId=10002&krypto=QJrbAudPd0vzXUGByeatog%3D%3D&ddkey=http:ProductDisplay.

9. Marc Gunther, "Eco-Police Find New Target: Oreos," *Money*, August 21, 2008. Available online at http://money.cnn.com/2008/08/21/news/companies/palm_oil.fortune/index.htm?postversion=2008082112

10. Ibid.

11. USDA FAS, "Indonesia: Palm Oil Production Prospects Continue to Grow," December 31, 2007. Total area for Indonesia palm oil in 2006 is estimated at 6.07 million hectares according to information from the Indonesia Palm Oil Board (IPOB). Available online at http://www.pecad.fas.usda.gov/highlights/2007/12/Indonesia_palmoil/.

12. "New Data Analysis Conclusive About Release of $CO_2$ When Natural Swamp Forest Is Converted to Oil Palm Plantation," CARBOPEAT Press Release, December 3, 2007. Dr. Sue Page or Dr, Chris Banks (CARBOPEAT Project Office), Department of Geography, University of Leicester, UK.

13. USDA FAS.

14. "Palm Oil Firm Wilmar Harming Indonesia Forests-Group," Reuters, July 3, 2007. Available at http://www.alertnet.org/thenews/newsdesk/SIN344348.htm.

15. Bunge Corporate Website. Online at http://www.bunge.com/about-bunge/promoting_sustainability.html.

16. See information at Cargill-Malaysia's website, http://www.cargill.com.my/, and Cargill-Indonesia, http://www.cargill.com/news/issues/palm_current.htm.

17. See, for instance, Cargill's position statement: http://www.cargill.com/news/issues/palm_roundtable.htm#TopOfPage. Bunge: http://www.bunge.com/about-bunge/promoting_sustainability.html.

18. Greenpeace. See, for instance, http://www.greenpeace.org.uk/forests/faq-palm-oil-forests-and-climate-change.

19. "New Data Analysis . . ." For more information, see "Carbon–Climate–Human Interactions in Tropical Peatlands: Vulnerabilities, Risks & Mitigation Measures."

20. Steinfeld et al., xxi.

21. Ibid., xxi.

22. Ibid.

23. British Government Panel on Sustainable Development, *Third Report*, 1997. Department of the Environment.

24. From company annual reports, Tyson and Smithfield, 2007.

25. Steinfeld et al., 45.

26. For further discussion, see Paul Roberts, *The End of Food* (Boston: Houghton Mifflin, 2008), 293. See also Frances Moore Lappé, *Diet for a Small Planet*, 20th anniversary ed. (New York: Ballantine Books, 1991).

27. Conversion ratios from USDA, from Allen Baker, Feed Situation and Outlook staff, ERS, USDA, Washington, D.C.

28. Roberts, quoting "Legume versus Fertilizer Sources of Nitrogen: Ecological Trade-offs and Human Need," *Agriculture, Ecosystems, and Environment* 102 (2004): 293.

29. World GHG Emissions Flow Chart, World Resources Institute, Washington, D.C. Based on data from 2000. All calculations are based on $CO_2$ equivalents, using 100-year global warming potentials from the IPCC (1996). Land use change includes both emissions and absorptions. Available online at http://cait.wri.org/figures.php?page=/World-FlowChart.

30. According to the IPCC, greenhouse gases relevant to radiative forcing include the following (parts per million [ppm] and parts per trillion [ppt] are based on 1998 levels): carbon dioxide ($CO_2$), 365 ppm; methane ($CH_4$), 1,745 ppb; nitrous oxide ($N_2O$), 314 ppb; tetrafluoromethane ($CF_4$), 80 ppt; hexafluoromethane ($C_2F_6$), 3 ppt; sulfur hexafluoride ($SF_6$), 4.2 ppt; trifluoromethane ($CHF_3$), 14 ppt; 1, 1, 1, 2-tetrafluoroethane ($C_2H_2F_4$), 7.5ppt; 1,1-Difluoroethane ($C_2H_4F_2$), 0.5ppt.

31. IPCC, *Climate Change 2007: Fourth Assessment Report of the Intergovernmental Panel on Climate Change* (New York: Cambridge University Press, 2007). Graphic 13.5.

32. World GHG Emissions Flow Chart, World Resources Institute.

33. Steinfeld et al., 79. See also, for instance, http://www.fao.org/ag/magazine/0612sp1.htm.

34. See, for example, Carbon Farmers of Australia. http://www.carbonfarmersofaustralia.com.au.

35. Steinfeld et al.

36. United Nations FAO, quoting Anthony Weis, *The Global Food Economy: The Battle for the Future of Farming* (London: Zed Books, 2007), 19.

37. J. McMichael et al., "Food, Livestock Production, Energy, Climate Change, and Health," *The Lancet* 370 (2007): 1253–1263.

38. Pachauri.

39. Steinfeld et al.

40. Ibid.

41. CNN, "All About: Food and Fossil Fuels," March 17, 2008, cnn.com. Available online at http://edition.cnn.com/2008/WORLD/asiapcf/03/16/eco.food.miles/; author communication with Professor Jonathan Lynch, University of Pennsylvania.

42. Author communication with Lynch.

43. Stern.

44. See for instance, Niles Eldredge, Life on Earth: An Encyclopedia of Biodiversity, Ecology, and Evolution (Santa Barbara, Calif.: ABC-CLIO, 2002). Online at http://www.landinstitute.org/vnews/display.v/ART/2002/08/23/439bd36c9acf1.

45. World GHG Emissions Flow Chart, World Resources Institute.

46. For more detail, see Environmental Protection Agency, "General Information on the Link Between Solid Waste and Greenhouse Gas Emissions." Available online at http://www.epa.gov/climatechange/wycd/waste/generalinfo.html#q1.

47. IPCC. See Figure 1, Chapter 2.

48. Most recent data available from USDA/ERS, U.S. Cattle and Beef Industry, 2002–2007. Available online at http://www.ers.usda.gov/news/BSECoverage.htm.

49. Pounds noted here are measured by commercial carcass weight. U.S. Red Meat and Poultry Forecasts. Source: World Agricultural Supply and Demand Estimates and Supporting Materials. From USDA/ERS. See also http://www.ers.usda.gov/Browse/TradeInternationalMarkets/.

50. Data from Brazilian Beef Industry and Exporters Association. Cited in "Brazilian Beef Break Records in September," October 3, 2008, The Beef Site. Available online at http://www.thebeefsite.com/news/24565/brazilian-beef-break-records-in-september.

51. IPCC.

52. http://www.rodaleinstitute.org.

53. See, for instance, studies from the Rodale Institute, found here: http://www.newfarm.org/depts/NFfield_trials/1003/carbonsequest.shtml.

54. Editorial, "Deserting the Hungry?" *Nature* 451 (17 January 2008): 223–224 | doi:10.1038/451223b; published online January 16, 2008. Available at http://www.nature.com/nature/journal/v451/n7176/full/451223b.html.

55. Executive Summary, 9. IAASTD, "Summary Report," paper presented at the International Assessment of Agricultural Science and Technology for Development, Johannesburg, South Africa, April 2008.

56. "Civil Society Statement from Johannesburg, South Africa: A New Era of Agriculture Begins Today," April 12, 2008. Available online at http://www.agassessment.org/docs/Civil_Society_Statement_on_IAASTD-28Apr08.pdf.

57. Greenpeace Press Release, "Urgent Changes Needed in Global Farming Practices to Avoid Environmental Destruction," April 15, 2008.

58. Fifty-seven governments approved the Executive Summary of the Synthesis Report. An additional three governments—Australia, Canada, and the United States of America—did not fully approve the Executive Summary of the Synthesis Report, and their reservations are entered in the Annex. From the Executive Summary of IAASTD, "Summary Report."

59. *Nature*, 223–224.

60. Author interview with Martin Clough, head of biotech R & D and president of Syngenta Biotechnology, Inc., based in North Carolina; and Anne Birch, director with Corporate Affairs, Syngenta, September 9, 2008.

61. Nadia El-Hage Scialabba and Caroline Hattam, "General Concepts and Issues in Organic Agriculture," in *Organic Agriculture, Environment and Food Security*, ed. Environment and Natural Resources Service Sustainable Development Department (Rome: Food and Agriculture Organization of the United Nations, 2002), chapter 1. Available online at http://www.fao.org/docrep/005/y4137e/y4137e01.htm#P0_3.

62. "Organic crops perform up to 100 percent better in drought and flood years," November 7, 2003, Rodale Institute. Online at www.newfarm.org.

63. D. G. Hole et al., "Does Organic Farming Benefit Biodiversity?," Biological Conservation 122 (2005): 113–130, quoting James Randerson, "Organic Farming Boosts Biodiversity," *New Scientist* October 11, 2004. Note: *New Scientist* emphasizes that neither of the two groups of researchers—from the government agency, English

Nature, and from the Royal Society for the Protection of Birds—"has a vested interest in organic farming."

64. Jules Pretty, *Agroecological Approaches to Agricultural Development* (Essex: University of Essex, 2006).

65. Ibid.

## ANOTHER TAKE: GLOBAL WARMING AND YOUR FOOD

1. David Pimentel, L. Armstrong, C. Flass, F. Hopf, R. Landy, and M. Pimentel, *Interdependence of Food and Natural Resources in Food and Natural Resources*, ed. David Pimentel and Carl Hall (San Diego: Academic Press, 1989). See also David Pimentel and Mario Giampietro, *Food, Land, Population, and the U.S. Economy*, Carrying Capacity Network, 1994. Available online at http://www.dieoff.com; Hunter L. Lovins, and Christopher Juniper, "Energy and Sustainable Agriculture," The John Pesek Colloquium on Sustainable Agriculture, March 9, 2005; David Pimentel, "Impacts of Organic Farming on the Efficiency of Energy Use in Agriculture," Organic Center, Cornell University, August 2006. My own calculations were made using data from: T. West and G. Marland, "A Synthesis of Carbon Sequestration, Carbon Emissions, and Net Carbon Flux in Agriculture: Comparing Tillage Practices in the United States," *Agriculture, Ecosystems, and Environment* 91 (2002): 217–232; General Accounting Office (GAO), "Agricultural Pesticides: Management Improvements Needed to Further Promote Integrated Pest Management," GAO-01-815, 2001; Food and Agriculture Organization of the United Nations (FAO UN), Statistical Yearbook. Available online at www.fao.org/statistics/yearbook/vol_1_1/pdf/a07.pdf.

2. Luise Giani and Elke Ahrensfeld, "Pedobiochemical Indicators for Eutrophication and the Development of 'Black Spots' in Tidal Flat Soils in the North Sea Coast," Journal of Plant Nutrition and Soil Science 165 (2002): 537–543.

3. T. C. Daniel, A. N. Sharpley, D. R. Edwards, R. Wedepohl, and J. L. Lemunyon, "Minimizing Surface Water Eutrophication from Agriculture by Phosphorous Management," *Journal of Soil and Water Conservation* 49 (1994): 30–38; A. N. Sharpley, S. C. Chapra, R. Wedepohl, J. T. Sims, T. C. Daniel, and K. R. Reddy, "Managing Agricultural Phosphorous for Protection of Surface Waters: Issues and Options," *Journal of Environmental Quality* 23 (1994): 437–451; National Centers for Coastal Ocean Science (NCCOS), "Hypoxia in the Gulf of Mexico: Progress Towards the Completion of an Integrated Assessment." Available online at http://oceanservice .noaa.gov/products/pubs_hypox.html; United States Department of Agriculture (USDA), Census of Agriculture. National Agricultural Statistics Service, 2002.

4. The term "environmental foodprint" was coined by Jennifer Wilkins of Cornell University's Division of Nutritional Sciences in 2006. The Cool Foods Campaign uses the term "FoodPrint" to refer to an individual's contribution to global warming, based upon the food they eat. This includes the total amount of greenhouse gases produced to grow, process, package, and transport that food.

5. U.S. EPA, *Inventory of U.S. Greenhouse Gas Emissions and Sinks: 1990–2005* (Washington, D.C.: Environmental Protection Agency, 2007).

6. Elke Gianiand Ahrensfeld.

7. D. L. Phillips, D. White, and B. Johnson, "Implications of Climate Change Scenarios for Soil Erosion Potential in the USA," *Land Degradation and Rehabilitation* 4 (1993): 61–72.

8. R. Lal, "Soil Carbon Sequestration Impacts on Global Climate Change and Food Security," Science 304 (2004): 1623–1627; USDA, "Agricultural Chemical Usage: 1999 Cattle and Cattle Facilities," National Agricultural Statistics Service, 2000. Available online at http://usda.mannlib.cornell.edu/usda/current/AgChemUsCa/AgChemUsCa-04-26-2000.pdf.

9. USDA, "Agricultural Chemical Usage: Swine and Swine Facilities," National Agricultural Statistics Service, 2006. Available online at http://usda.mannlib.cornell .edu/usda/current/AgChemUseSwine/AgC hemUseSwine-12-20-2006.pdf; Union of Concerned Scientists (UCS), "Hogging It: Estimates of Antimicrobial Abuse in Livestock," January 2001, 60. Available online at http://www.ucsusa.org/assets/documents/ foods_and_environment/ hog_chaps.pdf.

10. Martin C. Heller and Gregory A. Keoleian, "Life Cycle-Based Sustainability Indicators for Assessment of the U.S. Food System," Center for Sustainable Systems, University of Michigan, Report No. CSS00-04, December 6, 2000. Available online at http://www.public.iastate.edu/~brummer/papers/FoodSystemSustainability.pdf.

## CHAPTER SEVEN: CHEAP FOOD

1. Centers for Disease Control, "Heat-Related Deaths Among Crop Workers: United States, 1992–2006," 57, no. 24 (June 2008). Available online at http://www .cdc.gov/mmwr/preview/mmwrhtml/mm5724a1.ht.

2. Garance Burke, "More Farm Deaths in Heat Despite Calif. Crackdown," Associated Press, August 21, 2008. Available online at http://www.usatoday.com/ news/nation/2008–08–20–3205167992_x.htm.

3. "California: Strawberries, Vegetables, Water" *Rural Migration News* 14, no. 3 (July 2008). Available online at http://migration.ucdavis.edu/rmn/comments.php ?id=1330_0_5_0.

4. William Kandel, "A Profile of Hired Farmworkers, a 2008 Update," Economic Research Report No. ERR-60, USDA, July 2008. Available online at http://www .ers.usda.gov/publications/err60/err60.pdf.

5. U.S. General Accountability Office (GAO), "Pesticides: Improvement Needed to Ensure the Safety of Farm workers and Their Children," GAO/RCED-00-40, 2000, citing a 1993 U.S. EPA study. Available online at www.gao.gov/archive/2000/rc00 040.pdf.

6. Oxfam, "Like Machines in the Fields: Workers Without Rights in American Agriculture." Research Paper, March 2004. Available online at http://www.oxfam america.org/newsandpublications/publications/research_reports/art7011.html/ OA-Like_Machines_in_the_Fields.pdf.

7. Ibid., citing Linda Calvin et al., *U.S. Fresh Fruit and Vegetable Marketing: Emerging Trade Practices, Trends, and Issues* (Washington, D.C.: Economic Research Service, U.S. Department of Agriculture, January 2001, Agricultural Economic Report No. 795. Available online at http://www.ers.usda.gov/publications/aer795.

8. USDA, "Food CPI, Prices and Expenditures: Food Expenditures by Families and Individuals as a Share of Disposable Personal Income," Economic Research Service, June 17, 2008. Available online at http://www.ers.usda.gov/briefing/CPIFoodandExpenditures/Data/table7.htm.

9. Jane Black, "Slow Food at Full Speed: They Ate It Up." *Washington Post*, September 3, 2008. Available online at http://www.washingtonpost.com/wp-dyn/content/story/2008/09/02/ST2008090202273.html.

10. Hayden Stewart, "How Low Has the Farm Share of Retail Food Prices Really Fallen?" Economic Research Report No. ERR-24, August 2006. Available online at www.ers.usda.gov/Publications/ERR24/.

11. Daniel Rothenberg, *With These Hands: The Hidden World of Migrant Farmworkers Today* (Berkeley: University of California Press, 1998), 97.

12. Philip Martin, "Labor Relations in California Agriculture," In *University of California Institute for Labor and Employment. The State of California Labor*, 2001. Available online at http://repositories.cdlib.org/ile/scl2001/Section7.

13. Ibid.

14. Keith Cunningham-Parmeter, "A Poisoned Field: Farm Workers, Pesticide Exposure, and Tort Recovery in an Era of Regulatory Failure," *New York University Review of Law & Social Change* 28: 431.

15. Ibid.

16. Margaret Reeves, Anne Katten, and Martha Guzman, "Fields of Poison," (Darby, Pa.: Diane Publishing, 2002), citing a study by P. K. Mills and S. Kwong, "Cancer Incidence in the United Farm Workers of America (UFW) 1987–1997," *American Journal of Industrial Medicine* 40 (2001): 596–603. Available online at http://www.ufw.org/white_papers/report.pdf.

17. Liquid Gold: A California Exhibition. An Exhibit by the Water Resources Center Archives, University of California at Berkeley. Online at http://www.lib.berkeley.edu/WRCA/exhibit.html.

18. The Environmental Justice Water Coalition, "Thirsty for Justice: A People's Blueprint for California Water," 2005. Available online at http://www.ejcw.org/Thirsty%20for%20Justice.pdf.

19. U.S. Department of Labor (DOL), *Findings of the National Agricultural Workers Survey (NAWS) 2001–2002: A Demographic and Employment Profile of United States Farmworkers* (Washington, D.C: U.S. Department of Labor, March 2005), Research Report No. 9. Available online at http://www.dol.gove/asp/programs/agworker/report_8.pdf.

20. Kandel.

21. Martin.

22. Public Citizen, "Down on the Farm: NAFTA's Seven-Years War on Farmers and Ranchers in the U.S., Canada and Mexico," June 2001. Available online at http://www.citizen.org/documents/ACFF2.PDF.

23. Ibid.

24. Giselle Henriques and Raj Patel, "Agricultural Trade Liberalization and Mexico," Food First, Policy Brief 7, 2003, available online at http://www.foodfirst.org/pubs/policy/pb7.pdf.

25. Philip Martin, "NAFTA and Mexico-US Migration," 2005. Available online at http://giannini.ucop.edu/Mex_USMigration.pdf.

26. Tracy Wilkinson, "Less Money Going to Mexico as US Economy Falters," *Los Angeles Times,* October 2, 2008. Available online at http://www.latimes.com/news/printedition/asection/la-fg-mexmoney2-2008oct02,0,2037607.story.

27. Connie de la Vega and Conchita Lozano, 2005, "Advocates Should Use Applicable International Standards to Address Violations of Undocumented Workers' Rights in the United States," *Hastings Race & Poverty Law Journal* 3, 35.

28. General Accountability Office, "Illegal Immigration: Border Crossing Deaths Have Doubled Since 1995; Border Patrol's Efforts to Prevent Deaths Have Not Been Fully Evaluated," GAO-06-770, August 2006. Available online at http://www.gao.gov/new.items/d06770.pdf.

29. Department of Health and Human Services, "Annual Update of the HHS Poverty Guidelines," *Federal Register* 73, no. 15 (January 23, 2008): 3971–3972. Available online at http://aspe.hhs.gov/POVERTY/08fedreg.htm.

30. Kandel.

31. Ibid.

32. DOL.

33. Kandel.

34. Ibid.

35. DOL.

36. Ibid.

37. Eduardo Porter, "Illegal Immigrants Are Bolstering Social Security with Billions," *New York Times*, April 5, 2005. Available online at www.nytimes.com/2005/04/05/business/05immigration.html.

38. Don Villarejo et al., "Suffering in Silence: A Report on the Health of California's Agricultural Workers," California Institute of Rural Studies, Sponsored by California Endowment, 2001. Available online at www.fachc.org/pdf/mig_suffering%20in%20silence.pdf.

39. Alina Tugend, "The Least Affordable Place to Live? Try Salinas," *New York Times,* May 7, 2006, real estate section. Available online at www.nytimes.com/2006/05/07/realestate/07california.html.

40. USDA, Census on Agriculture, 2002. Available online at http://www.nass.usda.gov/Census/Pull_Data_Census.jsp.

41. Data were gathered from OSHA's inspection database. Query included all establishments engaged in crop or livestock production from September 1, 2007, and September 1, 2008. Available online at http://www.osha.gov/pls/imis/industry.html. 994 inspections were conducted during the same time period for all crop and livestock establishments (approximately 2.1 million farms or 938 million acres) in the United States during the same time period.

42. *United Farm Workers v. VINCENT B. ZANINOVICH & SONS, A CALIFORNIA CORPORATION,* 34 ALRB No. 3 (2008).

43. Burke.

44. Robert Gordon, "Poisons in the Fields: The United Farm Workers, Pesticides, and Environmental Politics," *The Pacific Historical Review* 68, no. 1 (Feb. 1999): 51–77.

Available online at http://links.jstor.org/sici?sici=0030–8684%28199902%2968%3A 1%3C51%3APITFTU%3E2.0.CO%3B2–8.

45. "How We Eat: 2005," *Rural Migration News* 13, no. 3 (July 2007). Available online at http://migration.ucdavis.edu/rmn/more.php?id=1229_0_5_0.

46. Ibid.

47. Address by Cesar Chavez, President, United Farm Workers of America, AFL-CIO, Pacific Lutheran University, Tacoma, Washington, March 1989. Available online at http://www.ufw.org/_page.php?menu=research&inc=history/10.html.

## ANOTHER TAKE: FIELDS OF POISON

1. The Department of Pesticide Regulation is the primary regulatory agency responsible for enforcing federal and state worker safety laws. *Summary of Results from the California Pesticide Illness Surveillance Illness Program*, 2006. CA EPA, DPR, Feb. 2008. HS-1872.

## CHAPTER EIGHT: THE FINANCIAL CRISIS AND WORLD HUNGER

1. "The New Face of Hunger," *The Economist*, April 17, 2008. Available online at http://www.economist.com/world/international/PrinterFriendly.cfm?story_id=110 49284.

2. Kate Smith and Rob Edwards, "2008: The Year of the Global Food Crisis," *Sunday Herald*, n.d. Available online at http://www.sundayherald.com/mostpopular .var.2104849.mostviewed.2008_the_year_of_global_food_crisis.php.

3. "New Face of Hunger."

4. "New Production System Needed to Tackle Global Food Crisis, Says UN Expert," United Nations Human Rights Council, October 16, 2008. Available online at http://www.reliefweb.int/rw/rwb.nsf/db900SID/SHIG-7KGFM5?OpenDocument.

## ANOTHER TAKE: WORLD HUNGER—YOUR ACTIONS MATTER

1. "Number of Hungry People Worldwide Nears 1 Billion Mark Agency Reports." UN News Centre. http://www.un.org/apps/news/story.asp?NewsID=29231 &Cr=Food+crisis&Cr1=.

2. Bread for the World homepage 2007. www.bread.org. March 7, 2008.

3. "Feeding America: Hunger and Poverty," statistics p2008. www.feedingamerica .org. December 10, 2008.

4. John Blake, "Charities Forced to Do More with Less," CNN.com. April 18, 2008. http://www.cnn.com/2008/LIVING/04/22/charity.shortage/index.html.

## CHAPTER TWELVE: IMPROVING KIDS' NUTRITION

1. R. Anderson, ""Deaths: Leading Causes for 2000," *National Vital Statistics Reports* 50, no. 16. Accessed January 21, 2003. Available online at http://www.cdc.gov/ nchs/data/nvsr/nvsr50/nvsr50_16.pdf.

2. D. Freedman, W. Dietz, S. Srinivasan, and G. Berenson, "The Relation of Overweight to Cardiovascular Risk Factors Among Children and Adolescents: The Bogalusa Heart Study," *Pediatrics* 103, no. 6 (1999): 1175–1182.

3. Pathobiological Determinants of Atherosclerosis in Youth (PDAY) Research Group, "Natural History of Aortic and Coronary Atherosclerotic Lesions in Youth; Findings from the PDAY Study," *Arteriosclerosis and Thrombosis* 13 (1993): 1291–1298.

4. O. Pinhas-Hamiel, L. Dolan, S. Daniels, D. Standiford, P. Khoury, and P. Zeitler, "Increased Incidence of Non-Insulin-Dependent Diabetes Mellitus Among Adolescents," *The Journal of Pediatrics* 128, no. 5 (1996): 608–615.

5. C. Ogden, K. Flegal, M. Carroll, and C. Johnson, "Prevalence and Trends in Overweight Among U.S. Children and Adolescents, 1999–2000," *Journal of the American Medical Association* 288, no. 14 (2002): 1728–1732.

6. M. Serdula, D. Ivery, R. Coates, D. Freedman, D. Williamson, and T. Byers, "Do Obese Children Become Obese Adults? A Review of the Literature," *Preventive Medicine* 22 (1993): 167–177.

7. U.S. Department of Health and Human Services, The Surgeon General's Call to Action to Prevent and Decrease Overweight and Obesity (Rockville, Md.: U.S. Department of Health and Human Services, Public Health Service, Office of the Surgeon General; 2001); R. Strauss, "Childhood Obesity and Self-Esteem," *Pediatrics* 105, no. 1 (2000): e15.

8. G. Wang, and W. Dietz, "Economic Burden of Obesity in Youths Aged 6 to 17 Years: 1979–1999," *Pediatrics* 109 (2002): e81.

9. U.S. Department of Agriculture, Office of Analysis, Nutrition, and Evaluation, *Changes in Children's Diets: 1989–1991 to 1994–1996*, Report No. CN-01-CD1 (Washington, D.C.: USDA, January 2001); Institute of Medicine, National Academies. *Dietary Reference Intakes: Energy, Carbohydrate, Fiber, Fat, Fatty Acids, Cholesterol, Protein, Amino Acids* (Washington, D.C.: National Academies Press, 2002).

10. K. Munoz, S. Krebs-Smith, R. Ballard-Barbash, and L. Cleveland, "Food Intakes of U.S. Children and Adolescents Compared with Recommendations," *Pediatrics* 100 (1997): 323–329 (erratum in *Pediatrics* 101 [1998]: 952–953).

11. L. Kann et al., "Youth Risk Behavior Surveillance—United States, 1999," *Morbidity and Mortality Weekly Report* 49, no. SS-5 (2000): 1–96.

12. Agricultural Research Service, US Department of Agriculture, *Food and Nutrient Intakes by Children 1994–96, 1998 (1999)*, Table Set 17. Accessed August 17, 2001. Available online at http://www.barc.usda.gov/bhnrc/foodsurvey/home.htm.

13. U.S. Department of Agriculture, Economic Research Service. *Per Capita Food Consumption Data System*, (Washington, D.C.: USDA). Accessed April 16, 2002. Available online at http://www.ers.usda.gov/Data/FoodConsumption/Spreadsheets/beverage.xls.

14. D. S. Ludwig, K. E. Peterson, and S. L. Gortmaker, "Relation Between Consumption of Sugar-Sweetened Drinks and Childhood Obesity: A Prospective, Observational Analysis," *The Lancet* 357 (2001): 505–508.

15. C. Ballew et al., "Beverage Choices Affect Adequacy of Children's Nutrient Intakes," *Archives of Pediatric and Adolescent Medicine* 154 (2000): 1148–1152; S. A. Bowman, "Diets of Individuals Based on Energy Intakes from Added Sugars," *Family Economics and Nutrition Review* 12 (1999): 31–38; P. M. Guenther, "Beverages in the

Diets of American Teenagers," *Journal of the American Dietetic Association* 86 (1986): 493–499; C. J. Lewis et al., "Nutrient Intakes and Body Weights of Persons Consuming High and Moderate Levels of Added Sugars," *Journal of the American Dietetic Association* 92 (1992): 708–713; L. Harnack et al., "Soft Drink Consumption among US Children and Adolescents: Nutritional Consequences," *Journal of the American Dietetic Association* 99 (1999): 436–441.

16. M. Jacobson, *Liquid Candy: How Soft Drinks Are Harming Americans' Health* (Washington, D.C.: Center for Science in the Public Interest, 1998).

17. "National School Lunch Program, General Purpose and Scope." Federal Register: 7CFR § 210.1.

18. J. Dwyer, "The School Nutrition Dietary Assessment Study," *American Journal of Clinical Nutrition* 61 (1995): 173S–177S.

19. USDA, "The National School Lunch Program." Accessed November 16, 2008. Available online at http://www.fns.usda.gov/cnd/Lunch/AboutLunch/NSLPFactSheet.pdf.

20. USDA, "The School Breakfast Program." Accessed November 16, 2008. Available online at http://www.fns.usda.gov/cnd/Breakfast/AboutBFast/SBPFactSheet.pdf.

21. Centers for Disease Control and Prevention, *School Health Policies and Programs Study 2006*. Accessed August 5, 2008. Available online at http://www.cdc.gov/HealthyYouth/shpps/2006/factsheets/pdf/FS_Nutrition_SHPPS2006.pdf.

22. Food and Nutrition Service (FNS), U.S. Department of Agriculture, *School Nutrition Dietary Assessment Study—III* (Alexandria, Va.: FNS, 2007).

23. A. Subar, S. Krebs-Smith, A. Cook, and L. Kahle, "Dietary Sources of Nutrients Among U.S. Children, 1989–1991." *Pediatrics* 102 (1998): 913–923.

24. Government Accountability Office (GAO), "School Meal Programs: Competitive Foods Are Widely Available and Generate Substantial Revenues for Schools" (Washington, D.C.: GAO, August 2005).

25. FNS.

26. "Requirements for School Food Authority Participation, Competitive Food Services." Federal Register: 7 CFR § 210.11.

27. CSPI, *State School Foods Report Card 2007* (Washington, D.C.: CSPI, 2007).

28. Centers for Disease Control and Prevention, *School Health Policies and Programs Study 2000*. Accessed September 19, 2001. Available online at http://www.cdc.gov/nccdphp/dash/shpps/factsheets/fs00_ns.htm.

29. USDA, Foods Sold in Competition with USDA School Meal Programs: A Report to Congress, January 12, 2001 (Washington, D.C.: USDA, 2001).

30. Ibid.

31. H. Wechsler et al., "Food Service and Foods and Beverages Available at School: Results from the School Health Policies and Programs Study 2000," *Journal of School Health* 71 (2001): 313–324.

32. USDA, *Foods Sold.*

33. K. Munoz et al.

34. U.S. Department of Agriculture, Office of Analysis, Nutrition, and Evaluation, *Changes in Children's Diets: 1989–1991 to 1994–1996*. Report No. CN-01-CD1 (Washington, D.C.: USDA, January 2001).

35. Ibid.; Institute of Medicine.

36. L. Jahns et al., "The Increasing Prevalence of Snacking Among U.S. Children from 1977 to 1996," *The Journal of Pediatrics* 138 (2001): 493–498.

37. USDA, Office of Analysis, Nutrition, and Evaluation.

38. Mathematica Policy Research, *Children's Diets in the Mid-1990s. Dietary Intake and Its Relationship with School Meal Participation*, final report submitted to the U.S. Department of Agriculture (Princeton, N.J.: Mathematica, 2001); Mathematica Policy Research, *Children's Diets 1989–91 to 1994–96*, final report submitted to the U.S. Department of Agriculture (Princeton, N. J.: Mathematica, 2001).

39. Harnack et al.; P. M. Guenther, "Beverages in the Diets of American Teenagers," *Journal of the American Dietetic Association* 86 (1986): 493–499.

40. D. S. Ludwig et al., "Relation Between Consumption of Sugar-Sweetened Drinks and Childhood Obesity: A Prospective, Observational Analysis," *Lancet* 357 (2001): 505–508.

41. J. Schwimmer, T. Burwinkle, and J. Varni, "Health-Related Quality of Life of Severely Obese Children and Adolescents," *Journal of the American Medical Association* 289 (2003): 1813–1819.

42. C. Ballew et al., "Beverage Choices Affect Adequacy of Children's Nutrient Intakes," *Archives of Pediatric and Adolescent Medicine* 154 (2000): 1148–1152; S. A. Bowman, "Diets of Individuals Based on Energy Intakes from Added Sugars," *Family Economics and Nutrition Review* 12 (1999): 31–38; Guenther; C. J. Lewis et al., "Nutrient Intakes and Body Weights of Persons Consuming High and Moderate Levels of Added Sugars," *Journal of the American Dietetic Association* 92 (1992): 708–713; Harnack et al.

43. USDA, Office of Analysis, Nutrition, and Evaluation.

44. Jacobson.

45. National Osteoporosis Foundation, *Disease Statistics, Fast Facts*. Accessed August 16, 2002. Available online at http://www.nof.org/osteoporosis/stats.htm.

46. American Dental Association, Joint Report of the American Dental Association Council on Access, Prevention, and Interprofessional Relations and Council on Scientific Affairs to the House of Delegates: Response to Resolution 73H-2000. October 2001.

47. Food and Nutrition Service, U.S. Department of Agriculture; Centers for Disease Control and Prevention, U.S. Department of Health and Human Services; and U.S. Department of Education, *Making it Happen! School Nutrition Success Stories*. FNS-374 (Alexandria, Va.: U.S. Department of Education, January 2005).

48. Wes Clark, personal communication, September 27, 2006.

49. Center for Weight and Health, University of California at Berkeley, "Dollars and Sense: The Financial Impact of Selling Healthier School Foods." Available online at http://www.cnr.berkeley.edu/cwh/PDFs/Dollars_and_Sense_FINAL_3.07.pdf.

# INDEX

I believe that a good story well told can truly make a difference in how one sees the world. This is why I started Participant Media: to tell compelling, entertaining stories that create awareness of the real issues that shape our lives.

At Participant, we seek to entertain our audiences first, and then invite them to participate in making a difference. With each film, we create social action and advocacy programs that highlight the issues that resonate in the film and provide ways to transform the impact of the media experience into individual and community action.

Twenty films later, from GOOD NIGHT AND GOOD LUCK, to AN INCONVENIENT TRUTH, and from THE KITE RUNNER to THE SOLOIST, and through thousands of social action activities, Participant continues to create entertainment that inspires and compels social change. Now through our partnership with PublicAffairs, we are extending our mission so that more of you can join us in making our world a better place.

Jeff Skoll

PublicAffairs is a publishing house founded in 1997. It is a tribute to the standards, values, and flair of three persons who have served as mentors to countless reporters, writers, editors, and book people of all kinds, including me.

I. F. Stone, proprietor of *I. F. Stone's Weekly*, combined a commitment to the First Amendment with entrepreneurial zeal and reporting skill and became one of the great independent journalists in American history. At the age of eighty, Izzy published *The Trial of Socrates*, which was a national bestseller. He wrote the book after he taught himself ancient Greek.

Benjamin C. Bradlee was for nearly thirty years the charismatic editorial leader of *The Washington Post*. It was Ben who gave the *Post* the range and courage to pursue such historic issues as Watergate. He supported his reporters with a tenacity that made them fearless and it is no accident that so many became authors of influential, best-selling books.

Robert L. Bernstein, the chief executive of Random House for more than a quarter century, guided one of the nation's premier publishing houses. Bob was personally responsible for many books of political dissent and argument that challenged tyranny around the globe. He is also the founder and longtime chair of Human Rights Watch, one of the most respected human rights organizations in the world.

·    ·    ·

For fifty years, the banner of Public Affairs Press was carried by its owner Morris B. Schnapper, who published Gandhi, Nasser, Toynbee, Truman, and about 1,500 other authors. In 1983, Schnapper was described by *The Washington Post* as "a redoubtable gadfly." His legacy will endure in the books to come.

*Peter Osnos, Founder and Editor-at-Large*